International Cryogenics Monograph Series

The International Cryogenics Monograph Series was established in the early 1960s to present an opportunity for active researchers in various areas associated with cryogenic engineering to cover their area of expertise by thoroughly covering its past development and its present status. These high level reviews assist young researchers to initiate research programs of their own in these key areas of cryogenic engineering without an extensive search of literature.

More information about this series at http://www.springer.com/series/6086

Thomas J. Peterson · J. G. Weisend II

Cryogenic Safety

A Guide to Best Practice in the Lab and Workplace

Springer

Thomas J. Peterson
SLAC National Accelerator Laboratory
Menlo Park, CA, USA

J. G. Weisend II
European Spallation Source ESS, ERIC
Lund, Sweden

With Contributions by
John M. Jurns
Malmö, Sweden

Stephen S. Woods
Las Cruces, NM, USA

ISSN 0538-7051 ISSN 2199-3084 (electronic)
International Cryogenics Monograph Series
ISBN 978-3-030-16506-2 ISBN 978-3-030-16508-6 (eBook)
https://doi.org/10.1007/978-3-030-16508-6

Library of Congress Control Number: 2019936286

This Springer imprint is published by the registered company Springer Nature Switzerland AG
The registered company address is: Gewerbestrasse 11, 6330 Cham, Switzerland

Acknowledgements

The authors wish to thank all their colleagues in the cryogenic community that provided advice, suggestions and figures. Tom Spicer, Cindy Zitter and the rest of the Springer team have done their typically outstanding job of supporting the development of this book and its subsequent production. The authors are gratified that the book appears in the International Cryogenics Monograph Series. In addition:

John Weisend wishes to thank: his colleagues at the European Spallation Source and Lund University for their support and encouragement, his coauthors for their valuable contributions and as always his family: Shari, Rachel, Nick and Alex.

Tom Peterson thanks his colleagues in cryogenics and particle accelerator technology at Fermilab, SLAC, and in laboratories around the world for their support, advice, and our wonderfully collaborative work. My good fortune in my career was only possible with the support of my family, whom I thank for being who they are: Susan, Eric, and Karin.

Steve Woods wishes to thank: Dr. Donald Bruce Wilson, Chair of the New Mexico State University Chemical Engineering Department for exhorting him to pursue hydrogen endeavors and that hydrogen will be the energy currency of the future, Walter Stewart as a mentor and role model, and his colleagues at the NASA White Sands Test Facility for providing an exciting environment for learning about how hydrogen is used.

Introduction

The scope of cryogenics is vast. It is defined as the science and engineering of activities and systems whose temperature is lower than 120 K; cryogenics includes the separation of air into its constituent parts and the storage and transport of those parts, the use of liquid oxygen and liquid hydrogen as fuels for space exploration, the creation and transport of liquefied natural gas (LNG), the use of liquid helium for cooling superconducting magnets and RF cavities for use in particle accelerators and fusion reactors and for the cooling of infrared sensors. Cryogenics is also used to preserve cells and other biological specimens and is used in medical treatments. Cryogenic treatment of materials to enhance their performance or machinability is a growing area of development. Cryogenic facilities range in size from the Large Hadron Collider at CERN and large air separation plants to small top table devices in laboratories. See Figs. 1, 2, 3, 4 and 5.

People affected by cryogenic hazards range from experts in the field to technicians, truck drivers, students, plant workers and medical professionals who only use cryogenics peripherally during their day.

In all its guises, cryogenics presents unique safety hazards. These include issues associated with:

- The extreme cold of cryogenics
- Flammability of some cryogenic fluids (hydrogen, LNG)
- Enhanced combustion associated with the presence of liquid oxygen
- The displacement of oxygen by gases boiling off from cryogenic liquids (Oxygen Deficiency Hazards)
- The high pressures that can be formed during the volume expansion that occurs when a cryogenic fluid becomes a room temperature gas.

An additional challenge rises from the behavior of materials at cryogenic temperatures. Many materials are inappropriate for use in cryogenics and can fail, resulting in hazardous conditions. Make no mistake, cryogenics can be hazardous and injury and death have occurred in the field of cryogenics. People affected by cryogenic hazards range from experts in the field to those who only use cryogenics peripherally during their day. Engineers, students, technicians, truck drivers,

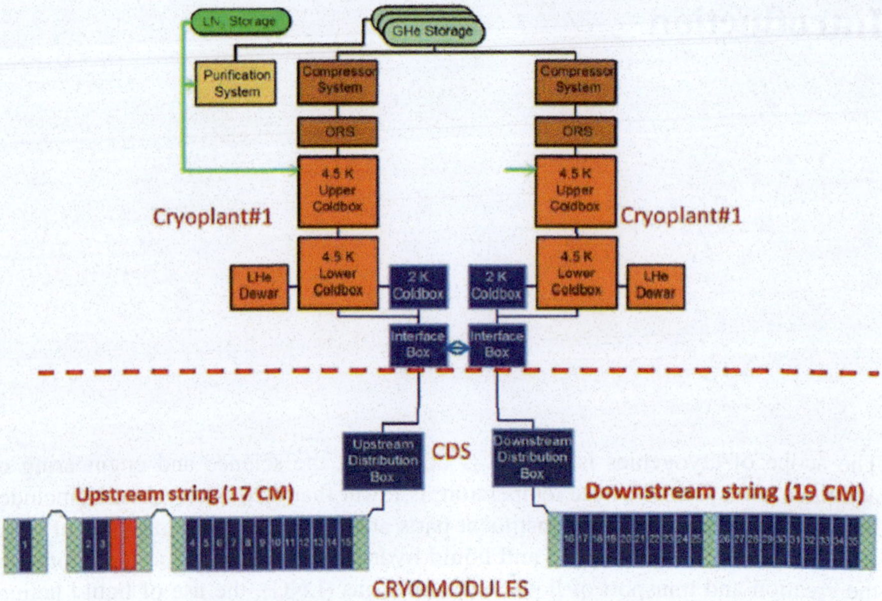

Fig. 1 Schematic of the LCLS II cryogenic system [1]

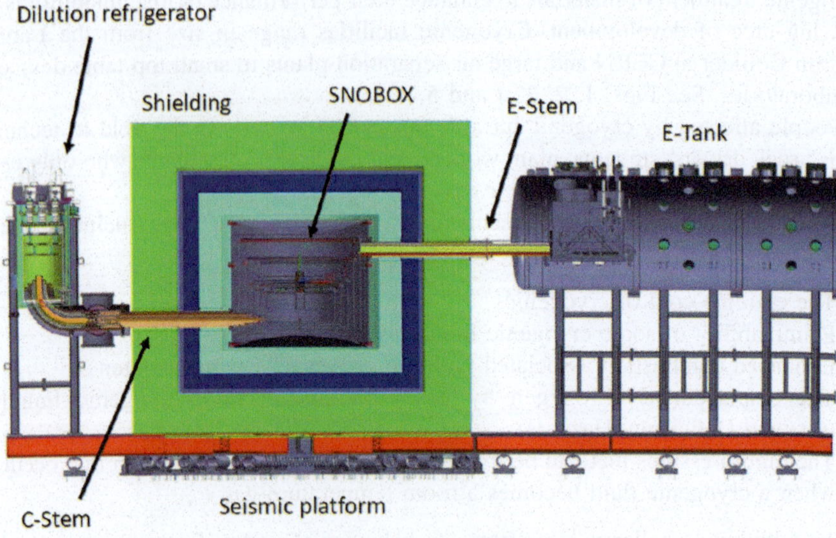

Fig. 2 Layout of the SuperCDMS SNOLAB experiment [2]

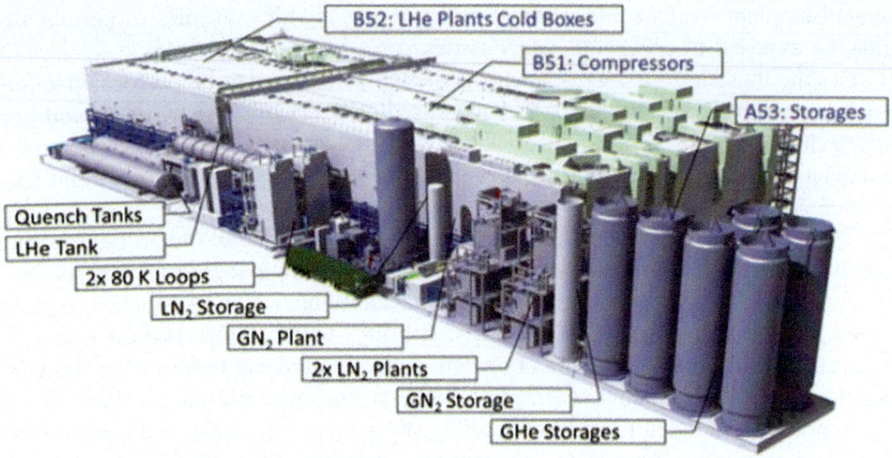

Fig. 3 Layout of ITER cryogenic buildings [3]

Fig. 4 Large pulse tube cooler (LPTC) developed jointly by CEA-SBT, Thales and Air Liquide for space applications [4]

Fig. 5 Liquid hydrogen turbopump for a cryogenic rocket engine [5]

scientists, plant workers and medical professionals are all examples of people who may be exposed to cryogenic safety issues.

Despite these hazards, work at cryogenic temperatures can be performed safely. Cryogenic safety is well understood and based on years of experience. In addition, research is ongoing as the field of cryogenics advances. Research on aspects of cryogenic safety is regularly presented at various cryogenic engineering conferences.

The goal of this book is to bring together the current state of the art in cryogenic safety. It is meant to be a useful reference that will allow people to use cryogenics safely as well as a pointer to other cryogenic safety resources. As such it includes charts, tables, figures and best practices. It brings together information from previous texts, industrial and laboratory safety polices and recent research papers. Case studies, example problems, descriptions of best practices and an extensive list of references are included to add to the utility of the text. This book is designed to be useful to everyone affected by cryogenic hazards regardless of their expertise in cryogenics.

Important Note

This book has been written and reviewed with every attempt to be accurate. Nevertheless, mistakes and omissions may occur. Please keep in mind that even when using the tools such as this book, safety remains the responsibility of the user.

References

1. V. Ravindranath et al., in *Process Simulations for the LCLS II Cryogenic Systems*. IOP Conference Series: Materials Science and Engineering, vol. 278 (2017)
2. M. I. Hollister et al., in *The Cryogenics Design of the SuperCDMS SNOLAB Experiment*. IOP Conference Series: Materials Science and Engineering, vol. 278 (2017)
3. E. Fauve et al., in *ITER Cryogenic Plant Infrastructures*. IOP Conference Series: Materials Science and Engineering, vol. 171 (2017)
4. L. Duband, Space Cryocooler Developments. *Phys. Procedia* **67** (2015)
5. R.S. Praveen, N. Jayan, K.S. Bijukumar, J. Jayaprakash, G. Ayyappan, in *Development of a Cryogenic Engine for GSLV MkIII:Technological Challenges*. IOP Conference Series: Materials Science and Engineering, vol. 171 (2017)

Contents

Authors and Contributors

About the Authors

Thomas J. Peterson has over 40 years of experience in large-scale cryogenic engineering for particle accelerators and test facilities. He holds a B.A. in mathematics and an M.S. in Engineering from the University of Wisconsin, Madison. His experience includes design, commissioning, and operation of large-scale helium cryogenic systems, including serving as project engineer for systems cooling superconductors in superfluid helium. Peterson's international experience includes 1.5 years at the Deutsches Electronen Synchrotron (DESY) in Hamburg, Germany and collaboration with CERN for Fermilab's contributions to the Large Hadron Collider (LHC). He joined the SLAC National Accelerator Laboratory in 2016 where his present position is Technical Director for the Linac Coherent Light Source II (LCLS-II) Project at SLAC.

J. G. Weisend II is currently Deputy Head of Accelerator Projects and Group Leader for Specialized Technical Services at The European Spallation Source, Lund, Sweden. He is also an Adjunct Professor at Lund University. He received his Ph.D. in Nuclear Engineering & Engineering Physics from the University of Wisconsin—Madison, where he investigated engineering applications of He II. He has worked at the SSC Laboratory, the Centre D'Etudes Nucleaires Grenoble, the Deutsches Elecktronen-Synchrotron Laboratory (DESY), the Stanford Linear Accelerator Laboratory (SLAC), the National Science Foundation and Michigan State University.

Dr. Weisend's research interests include He II and large scale accelerator cryogenics. He is the Chairman of the Board of Directors of the Cryogenic Society of America (CSA). He has led the CSA Short Course Program since 2001.

He is Chief Technical Editor of Advances in Cryogenic Engineering. In addition to co-authoring more than 70 technical papers, Dr. Weisend is the author of He is for Helium, the co-author (with N. Filina) of Cryogenic Two-Phase Flow and the editor of the Handbook of Cryogenic Engineering and of Cryostat Design. He writes a regular column "Cryo Bios" for the publication *Cold Facts* and is a member of both the Cryogenic Engineering Conference and International Cryogenic Engineering Conference Boards.

Contributors

John M. Jurns is a senior engineer with a B.S. from SUNY at Buffalo and a M.S. from Cleveland State University. His technical background focuses primarily on cryogenic system and component design, development, research, and testing. He has extensive experience in both private industry and government research facilities. He began his career at what is now Praxair, designing bulk cryogenic storage systems. After 10 years in industry performing design and integration work with cryogenic and high-pressure gas systems, he moved on to research and technology, working for 24 years at the NASA Glenn Research Center. At NASA, he was responsible for cryogenic system research program development, design, engineering investigations, build-up and operation of cryogenic test facilities. Following NASA, he worked 6 years at the European Spallation Source project in Sweden as the lead engineer responsible for the specification, procurement, and installation of a cryogenic helium refrigeration system. John currently works at the NIST Center for Neutron Research in Gaithersburg Maryland, responsible for design and operation of the cryogenic system infrastructure supporting NIST's research reactor. His specific areas of expertise include familiarity with cryogenic hydrogen, methane, oxygen, nitrogen and helium; designing research and prototype hardware; testing cryogenic systems and components; heat transfer analysis; aerospace mission analysis; and developing cooling systems for large scale particle accelerator facilities.

Mr. Stephen S. Woods During his 28 years at the NASA White Sands Test Facility, Mr. Woods has held positions of Physicist and Senior Scientist while performing aerospace propellant hazards research and systems hazard assessment. Among his accomplishments are hydrogen standards development, advancing hydrogen hazards assessment methodology, and hydrogen safety course development.

In addition to research papers, publications edited include:

- *Safe Use of Oxygen and Oxygen Systems* [ASTM MNL36, first edition],
- *Basic considerations for the safety of hydrogen systems* [ISO/PDTR 15916] and
- *Guide to Safety of Hydrogen and Hydrogen Systems* [ANSI/AIAA G-095-2017].

Current activities include:

- Chair of the American Institute for Aeronautics and Astronautics Hydrogen Committee on Standards,

- Liaison activities for NASA to the DOE Hydrogen Safety Panel, ISO Technical Committee 197 Hydrogen Technologies, and the National Fire Protection Association committee, Hydrogen Technologies (NFPA 2).

Presently, he is involved in work to characterize exposure hazards from hydrogen fire in microgravity spacecraft-environments and design efforts to reduce hazards in hydrogen appliances used in spacecraft. He is the lead instructor for NASA's hydrogen safety courses and has participated in hazard assessment of numerous aerospace and industry hydrogen systems.

Chapter 1
Properties of Fluids and Materials at Cryogenic Temperatures

Abstract Understanding the properties of cryogenic fluids and the cryogenic properties of materials is vital to the safe operation of cryogenic systems. This chapter describes the properties of typical fluids used in cryogenics including helium, nitrogen, oxygen and hydrogen as they relate to safety. Lists of suitable and unsuitable materials for cryogenic systems are given and material properties most linked to safety are discussed. Unique safety issues such as the impact of ionizing radiation on liquid nitrogen and the flammability hazards associated with charcoal adsorbers are also covered. References to sources of material properties are given. A list of best practices is included. Additional specific properties are provided in Chap. 5 (Oxygen Systems), Chap. 6 (Hydrogen Systems) and Chap. 7 (LNG).

1.1 Example Accident

On 20 October, 1944 the sewers of a mixed-use residential and industrial neighborhood of Cleveland, Ohio started to explode. The cause was a failure of a Liquefied Natural Gas (LNG) storage tank at the East Ohio Gas Company facility. The sudden failure of this tank via a crack in the side spilled approximately 1.1 million gallons (4163 m^3) of LNG which quickly ignited, causing fires and explosions. Some of the LNG and cold vapor ran into the sewer system where it also caused fires and explosion. Due to the fires, a second LNG tank also failed releasing an additional 500,000 gallons (1893 m^3) that in turn ignited. This accident resulted in 128 deaths, damage to more than 200 buildings (more than 80 of which were completely destroyed) and roughly 3600 people made homeless [1]. Figure 1.1 provides a sense of the destruction.

It is very much worth pointing out the Cleveland facility was one of the first ever to liquefy natural gas and store it in liquid form for use during peak demand. The Cleveland accident greatly set back the use of LNG as a storage medium in the United States for many years. The modern LNG industry has become much safer and detailed safety measures have been developed (see Chap. 7). One of these measures is that LNG plants are no longer located in crowded neighborhoods.

T. J. Peterson and J. G. Weisend II, *Cryogenic Safety*, International Cryogenics Monograph Series, https://doi.org/10.1007/978-3-030-16508-6_1

Fig. 1.1 Resulting damage from the 1944 Cleveland LNG Fire and explosion. Gas plant located in the upper half of the picture; note destroyed homes, factories and other buildings in the lower half [2]

A single cause of the tank failure was never officially identified [2]. Suspicion has fallen on the use of a 3.5% nickel low carbon steel for the tank wall. This material was known to be brittle but was judged suitable for use and had been used successfully on other LNG tanks. One of the recommendations of the resulting Bureau of Mines investigation [2] was that such material should not be used in the particular cylindrical tank design employed in Cleveland unless it could be established conclusively that the material choice was not a cause of the accident. The materials used in modern LNG systems include: austenitic stainless steel, aluminum, Invar (35NiFe) and 9% nickel steel, all of which retain their toughness and ductility at LNG temperatures.

Another issue identified in the accident report was the ability of the LNG and very cold natural gas vapor to flow into low-lying sewers or ducts spreading the flammable substance into the surrounding neighborhood. The report recommended adding dikes around all future tanks to contain spilled LNG and cold vapor. This had not been required previously because the assumption was that any spill would quickly warm up to room temperature and be dispersed in the surrounding air. Thus, there were two potential places in this accident where the incorrect understanding of fluid and material properties contributed to the accident.

1.2 Introduction

Many safety issues in cryogenics arise directly from the behavior of fluids and materials at cryogenic temperatures. Understanding these behaviors and knowing which materials can be used at cryogenic temperatures is vital to safety in cryogenics. This chapter discusses those material properties most relevant to safety issues and provides pointers to later chapters in which the issues are addressed. More general surveys of material and fluid properties at cryogenic temperatures may be found in [3–7].

1.3 Cryogenic Fluids

Table 1.1 lists the most common cryogenic fluids along with their boiling point at 1 bar and the volume ratio between the gas at standard temperature and pressure (300 K and 1 bar) and the corresponding liquid at 1 bar. Cryogenics is generally defined as the science and engineering of phenomena occurring below 120 K. This table also includes propane and ethane with higher boiling points due to their presence in Liquefied Natural Gas (LNG) .

Even the small amount of data in Table 1.1 has significant implications for safety in cryogenics. These include possible pressurization or displacement of oxygen due to the large volume ratio.

Table 1.1 The boiling points and volume ratios of cryogenic fluids

Fluid	Normal boiling point (K)	Density of liquid at normal boiling point (kg/m^3)	Density of gas at 1 bar and 300 K (kg/m^3)	Volume of gas at 1 bar, 300 K/volume of liquid at normal boiling point
Propane	231.07	580.89	1.80	323
Ethane	184.55	543.97	1.22	446
Xenon	165.04	2942.1	5.29	556
Krypton	119.77	2416.3	3.40	711
Methane	111.63	422.42	0.64	660
Argon	87.28	1395.5	1.62	861
Oxygen	90.19	1142.2	1.3	879
Nitrogen	77.2	807.3	1.12	720
Neon	27.09	1205.2	0.81	1488
Hydrogen (para)	20.23	70.85	0.081	875
Helium	4.222	125.2	0.16	783

1.3.1 *Volume Ratios*

Table 1.1 shows the large volume ratio (a factor of 500 or more) between a cryogenic fluid, excluding propane and ethane, at its atmospheric boiling point and the volume of its corresponding gas at room temperature and pressure. This fact is the basis of two of the most significant hazards in cryogenics.

First, if a cryogenic fluid (or cryogen) is warmed up and converted to gas in an enclosed space very high pressures will be created which can lead to material failures and subsequent injury or death. Designing cryogenic systems that can safety vent such large volume expansions is the covered in Chap. 3.

Second, all the fluids except oxygen (which has its own hazards) will displace the breathable air in any area in which there are released. This causes an Oxygen Deficiency Hazard (ODH) in which the area can no longer support human life. ODH is one of the hazards in cryogenics most likely to lead to fatalities and must always be accounted for in the design and operation of cryogenic systems. Understanding and mitigating ODH hazards is described in Chap. 4. One of the issues to consider with ODH is the relative density of typical gases at room temperature and pressure. The density of dry air at 293 K and 1 bar is 1.204 kg/m^3. Examining Table 1.1 shows that the density of helium at these conditions is less than this value while the density of argon is greater than this value. The result is that undetected leaks of helium may result in oxygen deficient atmospheres near the ceiling while undetected leaks of argon may result in oxygen deficient atmospheres in trenches, pits or basements. This can result in danger to workers entering these areas.

Additionally, as demonstrated by the fire in Cleveland, during certain scenarios the venting or leaking cryogenic fluid may remain cold and dense for a significant time. This cold dense fluid may not mix well with the air and may accumulate in low-lying spaces. The cold leaking or venting fluid will also condense the water vapor in the air quickly, forming dense clouds that may inhibit evacuation. Consideration of the formation of clouds and density of the cold fluid must be considered when thinking about possible accidents and their mitigations.

1.3.2 *Flammability Hazards*

Propane, ethane, methane and hydrogen are flammable. The most common use of hydrocarbons in cryogenics is Liquefied Natural Gas (LNG) . This is generally composed if 95% methane, a few percent ethane and much smaller concentrations of propane and butane. Safety aspects of LNG are covered in Chap. 7.

There are many applications of liquid hydrogen (LH_2) including: use as a rocket fuel, scientific applications such as use as neutron moderator or as a target material in fundamental physics research. Liquid hydrogen is increasingly used as a terrestrial fuel. Another feature of hydrogen is that it exists in two spin states; ortho and para. At room temperature and pressure hydrogen is a mixture of ortho and para (known as "normal hydrogen"). At liquid hydrogen temperatures, the lowest energy state is parahydrogen and liquid hydrogen will naturally convert to this spin state but the process is slow and exothermic and thus it releases heat. This effect must be allowed for in any LH_2 system design. Additionally, the properties of normal, ortha and para hydrogen differ and care must be taken to use the properties of the correct spin state in calculations. The safe use of liquid hydrogen is discussed in Chap. 6.

1.3.3 Oxygen Hazards

While not itself a fuel, oxygen of course enables combustion and its presence greatly increase the risk of fire. Oxygen has many applications and is a major product of the air separation industry. Safe use of oxygen is described in detail in Chap. 5.

One feature of oxygen is that its boiling point at 1 bar is 90.1 K while that of nitrogen is 77.4 K. Since normal air is roughly 78% nitrogen and 21% oxygen, any exposed surface cooled to liquid nitrogen temperatures will condense air and this condensate will be oxygen rich due to the relative boiling points of oxygen and nitrogen. This oxygen rich condensate can pose a significant fire hazard. See Chap. 5 (Oxygen Safety) for an example calculation of this effect. Thus, it is better to always insulate cold surfaces so that this condensation does not occur. If this is not possible, then drip trays or other methods should be employed to manage the condensate and avoid creating a fire hazard.

A similar hazard can arise when using charcoal adsorbers cooled to liquid nitrogen temperatures in cryogenic plants or gas purification systems. If the flow stream moving through the adsorber has a sufficient oxygen content, the oxygen can condense of the charcoal thus mixing both a fuel (charcoal) and oxidizer (oxygen) together. Under these cases explosions can and have occurred. It may well be better to use non-combustible adsorbing material such as silica gel or molecule sieves in applications where significant oxygen content may be present [8].

1.3.4 Liquid Nitrogen and Ionizing Radiation

Liquid nitrogen can contain oxygen impurities either through production or through condensation of air into the LN_2 during transport and use. Ionizing radiation can convert some of this oxygen (O_2) into ozone (O_3) The ozone can then convert back to O_2 releasing enough energy to cause an explosion. Explosions have been observed in LN_2 systems exposed to large amounts of ionizing radiation. Examples have been observed with gamma rays, neutrons and high energy electrons [9–11]. The details of this phenomena are not completely understood and there may also be some additional contributions due to the formation of various nitrogen-oxygen compounds in the irradiated system. The radiation dose required for this phenomenon to occur is generally quite high, but there have been cases reported at lower doses as well [12].

Given the uncertainty surrounding this hazard, it is best not to subject LN_2 systems (and also liquid oxygen or liquid air systems) to ionizing radiation. At the very least, this hazard must be considered and the risk determined prior to operation.

An additional reason for avoiding the use of nitrogen in accelerator tunnels (where it might be subject to ionizing radiation) is the oxygen deficiency hazard posed by the nitrogen (Chap. 5).

1.3.5 Cold Hazards

At the risk of stating the obvious, cryogenic fluids are extremely cold and as result can cause frostbite and burns as well as eye damage. Safe handling of cryogenic fluids is described in Chap. 2. Particular care should be paid to eye protection. *Always wear appropriate eye protection when handling any amount of cryogenic fluids.* It is frequently observed that people working with liquid nitrogen in laboratory environments can be quite cavalier about the risk, but even a small splash of liquid nitrogen into the eyes can cause blindness.

The cold of cryogenic fluids can also adversely affect materials (see below) resulting in equipment damage, failures and additional hazards. It is important to note that it is not just the liquids but also the cold gas that vents from cryogenic systems that causes the hazard. Depending on the flow rate and the cryogenic fluid involved, this cold gas can easily be below 100 K near the vent. Such cold gas may not only pose a risk to personnel but also may cool down nearby materials (for example polymers) not designed to operate at these temperature. Always ensure that relief valves and vent lines are not pointed to where personnel may be nor pointed in such a way as to cool material not designed for cryogenic temperatures.

1.4 Cryogenic Properties of Materials

The properties of solid materials change greatly with temperature between 300 K and cryogenic temperatures. This variation in material properties must always be allowed for in cryogenic design. In addition, there are some materials that are completely inappropriate to use at cryogenic temperatures due to their properties at these temperatures. The use of inappropriate materials or neglecting to take into account the variation of material properties with temperature is frequently an underlying cause of cryogenic failures and accidents.

This section lists materials appropriate and inappropriate for use in cryogenics and describes the impact of some material properties on cryogenic safety.

1.4.1 *Appropriate and Inappropriate Materials for Cryogenics*

Many materials will not function properly at cryogenic temperatures and can fail resulting in safety hazards. An important rule is to never use a material at cryogenic temperatures that has not previously been proven to work at those temperatures. Table 1.2 is a list of materials that are appropriate for use at cryogenic temperatures while Table 1.3 is a list of materials that should not be used at cryogenic temperatures.

Examining Table 1.2, one can see that many of the workhorse materials in engineering, e.g. stainless steel, aluminum and copper are suitable for use in cryogenics. Some nonstandard materials are also listed. Invar, which is a Iron/Nickel alloy is used in cryogenics due to its low thermal contraction. Niobium is a superconductor that appears as an alloy (NbTi and Nb_3Sn) in superconducting wires and as a pure metal in superconducting radiofrequency cavities. Niobium does become brittle at cryogenic temperatures but experience has shown that it can be used at these temperatures and can even meet the spirit of various pressure vessel codes [13, 14]. While the materials in Table 1.2 can be used at cryogenic temperatures, their properties will change greatly with temperature and this change needs to be taken into account. In addition, some of the materials in Table 1.2 while suitable for cryogenics may not be suitable for use with Oxygen systems. These include, under certain conditions, titanium and aluminum. See Chap. 5 for more details.

Table 1.2 Some materials suitable for use in cryogenics [15]

Austenitic stainless steels e.g. 304, 304L, 316, 321
Aluminum alloys e.g. 6061, 6063, 1100
Copper e.g. OFHC, ETP and phosphorous deoxidized
Brass
Fiber reinforced plastics such as G-10 and G-11
Teflon (depending on the application)
Niobium & Titanium (frequently used in superconducting RF systems)
Invar (Ni/Fe alloy)
Indium (used as an O ring material)
Kapton and Mylar (used in multilayer Insulation and as electrical insulation)
Quartz (used in windows)

Table 1.3 Some materials unsuitable for use in cryogenics [15]

Martensitic stainless steels—undergoes ductile to brittle transition when cooled down
Cast iron—becomes brittle
Carbon steels—becomes brittle
Rubber and most plastics (important exceptions are Kel-F and UHMW used as seats in cryogenic valves)

Turning to Table 1.3, one can see that a common problem with unsuitable materials is that they become brittle at cryogenic temperatures. Some designs will use carbon steels as the material for the 300 K vacuum vessel that surrounds the cryogenic components. This application can work, but care must be taken to ensure that the vacuum vessel is not cooled below the metal's ductile to brittle transition temperature due to a leak in a cryogenic line or an unexpected connection to a cryogenic component. A more conservative approach is to use stainless steel for the vacuum vessels.

1.4.2 Hydrogen Embrittlement

Many commonly used engineering materials can become brittle in the presence of hydrogen. This can affect the strength of the material and result in failures well below the nominal yield strength of the material. Particular consideration must be given to material selection and use in both room temperature and cryogenic hydrogen systems. See Chap. 6 for more details.

1.4.3 Thermal Contraction

Most materials used in cryogenics shrink when cooled from 300 K to cryogenic temperatures. This thermal contraction must be allowed for in cryogenic design [15] and may have significant safety implications. During cool down, such contraction can lead to unexpected links between warm and cold surfaces; perhaps resulting in a material operating below its allowable temperature. A structure that is rigidly fixed may undergo high stress upon cool down, possibly resulting in failure. This last effect is commonly seen in wires connecting cryogenic temperatures to room temperature; if freedom (such as loops in the wire) is not given in the wiring design for this contraction to take place, breakage of wire or wire connections is common.

Table 1.4 lists the integrated shrinkage found in materials commonly used in cryogenics between 300 and 4 K, while Fig. 1.2 shows the integrated thermal expansion (note the negative sign indicating contraction) of aluminum and stainless steel alloys. The solution is to be aware of this problem and to design systems that allow for this effect. For example, do not over constrain structural or wiring system. Allowing for thermal contraction from the beginning, improves both the safety and reliability of the system.

Table 1.4 Integrated thermal contraction for practical engineering materials between 300 and 100 K and between 100 and 4 K [15]

Material	ΔL/L (300 – 100 K)	ΔL/L (100 – 4 K)
Stainless steel	296×10^{-5}	35×10^{-5}
Copper	326×10^{-5}	44×10^{-5}
Aluminum	415×10^{-5}	47×10^{-5}
Iron	198×10^{-5}	18×10^{-5}
Invar	40×10^{-5}	–
Brass	340×10^{-5}	57×10^{-5}
Epoxy/fiberglass	279×10^{-5}	47×10^{-5}
Titanium	134×10^{-5}	17×10^{-5}

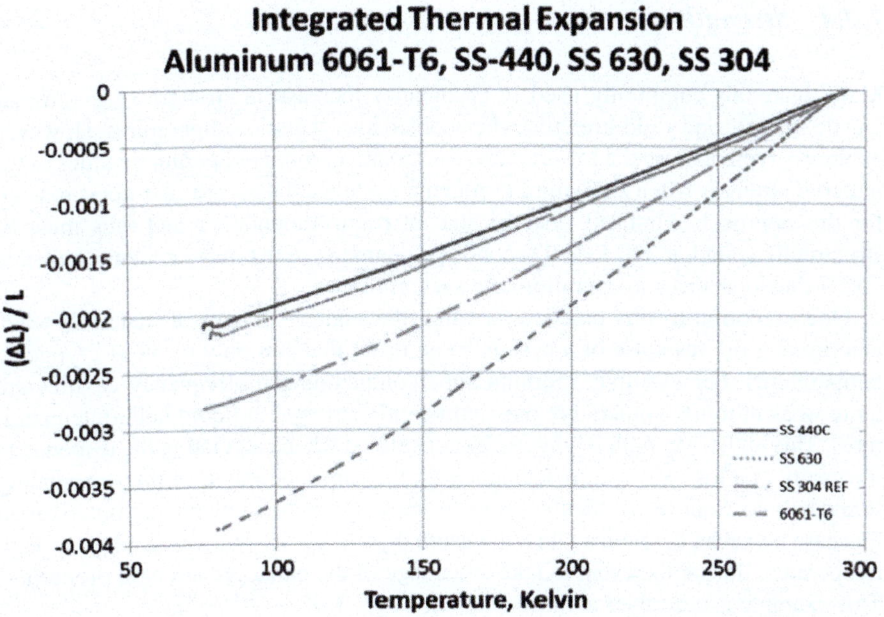

Fig. 1.2 Measured integrated thermal expansion for Al 606i-T6 and various stainless steel alloys [16]

Example Calculation

A 3.2 m length of copper wire runs parallel to and is connected at both ends to 3 m long length of stainless steel. Are there any issues if the combined stainless steel and copper assembly is cooled to 100 K?

Answer:

The stainless steel piece will contract by 3 m × 296 × 10^{-5} (Table 1.4) or 8.9 mm resulting in a new length of 3 – 0.0089 = 2.991 m

The copper will contract by 3.2 m × 326 × 10^{-5} (Table 1.4) or 10.4 mm resulting in a new length of 3.2 – 0.0104 = 3.186 m

Thus, the copper wire remains longer than the stainless steel piece and will not be overstressed, damaged or broken upon cool down. There should be no issues present.

1.4.4 Strength

Many materials commonly used in cryogenics increase in strength (e.g. ultimate tensile strength and yield strength) when cooled to cryogenic temperatures. However, engineers typically use the lower, more conservative, room temperature values of the material strength when designing cryogenic systems. There are two good reasons for this approach. First, all systems start at room temperature and thus must be structurally sound at 300 K and second, cryogenic systems may suddenly return to 300 K due to an error in operation or a system failure.

One exception to this practice, is using the higher strength of materials when determining the response of a system to an upset that can only occur at cryogenic temperatures. For example, niobium superconducting radiofrequency cavities are quite weak at room temperature but significantly stronger at liquid helium temperatures. This higher strength is sometimes considered when calculating the response of the cavity to a loss of vacuum accident with the subsequent boil of the surrounding liquid helium. Such an accident can only occur at liquid helium temperature. Even in this case however, care must taken to ensure that the cavity remains at a low enough temperature during the event to take advantage of the enhanced strength properties. This example is described in more detail in Chap. 3.

1.5 Sources of Material Property Data

1.5.1 Properties of Cryogenic Fluids

A very useful website (https://webbook.nist.gov/chemistry/fluid/) for calculating thermophysical properties of fluids, including those used in cryogenics, is maintained by the US National Bureau of Standards and Technology (NIST). This website includes properties in the cryogenic temperature range.

A set of commercial computer models that provide fluid property data are available from Horizon Technologies (http://www.htess.com/software.htm). These include:

- "HEPAK" provides helium property data including the superfluid helium range ($T < 2.2$ K).
- "GASPAK" provides property data for 33 different fluids.
- "He3PAK" provides property data for the ^{3}He isotope

All three of these programs are based on standards data from NIST.

Two other good sources of cryogenic fluid properties are: *Cryogenic Fluids Databook*, P. Cook and B.A. Hands, British Cryoengineering Society (2002) and *Thermodynamic Properties of Cryogenic Fluids*, J.W. Leachman, R.T. Jacobsen, E.W. Lemmon and S.G. Penoncello, Springer (2017).

1.5.2 Cryogenic Properties of Materials

The US National Bureau of Standards and Technology has a website (https://trc.nist.gov/cryogenics/materials/materialproperties.htm) with an index of material properties for materials including those used in cryogenics.

Horizon Technologies (http://www.htess.com/software.htm). Produces a number of computer codes that provide properties of materials at cryogenic temperatures. These include:

- "METALPAK" provides thermal property data for 14 metals
- "CPPAK" provides specific heat properties for materials including some not in METALPAK
- "EXPAK" provides thermal expansion properties materials including some not in METALPAK

1.5.3 Research on Material Properties

There is ongoing research on the properties of new and existing materials. A good source of this new information is found in the proceedings of a number of regular scientific conferences. Examples include:

(1) *Advances in Cryogenic Engineering*, Volumes 1–65, Plenum Press, AIP Press and IOP

These are the proceedings of the Cryogenic Engineering Conference/International Cryogenic Materials Conference which is held biannually (odd years) in North America.

Most recent conference (2017) available on line at
http://iopscience.iop.org/issue/1757-899X/278/1

(2) Proceedings of the International Cryogenic Engineering Conference

These are the proceedings of the International Cryogenic Engineering Conference that is held biannually (even years) in Europe or Asia
Most recent conference (2016) available online at:
http://iopscience.iop.org/issue/1757-899X/171/1

(3) Applied Superconductivity Conference, IEEE Transactions on Magnetics
(4) Proceedings of the Magnet Technology Conference

1.6 Best Practices

1. Only use materials at cryogenic temperatures that have been proven to operate at these temperatures
2. Ensure that the large variation of material properties with temperature has been taken into account for in the design
3. Always conduct an Oxygen Deficiency Hazard analysis (Chap. 4) when using cryogenic fluids or inert gases no matter how small the quantity involved.
4. Always take into account the volume expansion and subsequent pressure rise associated with cryogenic fluids. Design in appropriate pressure relief systems (Chap. 3)
5. Always wear appropriate personal protection equipment, including eye protection, when handling cryogenic fluids no matter how small the amount (Chap. 2).
6. Ensure that relief valves and vent lines do not direct the flow of cold gas towards people or towards materials not designed for cryogenic temperature use.
7. Avoid walking into or through visible vapor clouds.
8. Take into account the flammability hazards associated with Hydrogen (Chap. 6), LNG (Chap. 7) and other hydrocarbons.
9. Take into account the unique hazards associated with oxygen (Chap. 5)
10. Insulate lines and cold surfaces so that air does not condense on them. If this is not possible, install drip trays or other approaches to safely manage the resultant oxygen rich condensate.
11. Design for the significant thermal contraction that occurs upon cooling many materials down to cryogenic temperatures.

References

1. Cleveland suffers worst fire catastrophe in history. Fire Eng. **97**, 11 (1944)
2. Report on the investigation of the fire at the liquefaction, storage, and regasification plant of The East Ohio Gas Co., Cleveland, OH, 20 October, 1944, US Bureau of Mines, R.I. 3867 (1946)
3. R.P. Reed, A.F. Clark (eds.), *Materials at Low Temperatures*, American Society of Metals (1983)
4. J.G. Weisend II, V. Flynn, E. Thompson, R.P. Reed, A reference guide for cryogenic properties of materials, SLAC-TN-03-023, http://www.slac.stanford.edu/cgi-wrap/getdoc/slac-tn-03-023.pdf. Accessed 10 March 2017
5. T.F. Durham, R.M. McClintock, R.P. Reed, *Cryogenic Materials Data Handbook* (National Bureau of Standards, US Department of Commerce, 1961)
6. J.G. Weisend II (ed.), *The Handbook of Cryogenic Engineering* (Taylor & Francis, New York, 1998)
7. P. Cook, B.A. Hands, *Cryogenic Fluids Databook*, British Cryoengineering Society (2002)
8. Hazards associated with the use of activated carbon cryogenic gas purifiers, IGC Doc 43/14, European Industrial Gases Association (2014)
9. L. Bochirol, J. Doulat, L. Weil, A cryogenic device for irradiation in liquid nitrogen. Cryogenics (September, 1960)
10. C.R. Gregory, C.R. Nuttall, Explosion risks to cryogenic liquids exposed to ionizing radiation, CERN AT/95-06 (1995)
11. G.W. Chen, R.G. Struss, On the cause of explosions in reactor cryostats for liquid nitrogen. Cryogenics (April, 1969)
12. Explosion risk in liquid nitrogen, ATLAS AOS No. 12, CERN (2013)
13. T.J. Peterson, et al., Pure niobium as a pressure vessel material. Adv. Cryo. Engr. **55A** (2010)
14. T.J. Peterson, et al., A survey of pressure vessel code compliance methods for superconducting radio frequency cryomodules. Adv. Cryo. Engr. **57B** (2012)
15. J.G. Weisend II (ed.), *Cryostat Design* (Springer, 2016)
16. H. Cease et al., Cryogenic coefficient of thermal expansion measurements of type 440 and 630 stainless steels. Adv. Cryo. Engr. **60** (2014)

Chapter 2
General Cryogenic Safety

Abstract This chapter discusses aspects of cryogenic safety present in all cryogenic installations. Included are: the effects of cryogenic temperatures and the use of personal protective equipment to protect the body and eyes; first aid procedures for cryogenic burns; safe handling of dewars and cryostats; and hazards associated with system pressure and the presence of magnetic fields. A primary way to avoid accidents in cryogenic facilities is through the use of job hazard analysis and the development of a safe work culture. These topics are also covered. A list of best practices is included. It's important to note that this chapter is not a complete discussion of cryogenic safety issues. In particular Oxygen Deficiency Hazards (Chap. 4) and details of pressure relief systems (Chap. 3) are not covered here.

2.1 Example Accident

One day in the 1980s, an undergraduate student working in the University of Wisconsin Cryogenics Lab tried to save time. He had just completed an experiment in a relatively small vertical helium cryostat that included in its design an annular bath of liquid nitrogen (LN_2). The student was in a hurry to warm up the experiment. The helium in the central reservoir had already evaporated but the LN_2 reservoir was still 50% full and was taking a long time to evaporate. The student decided he could speed things up by taking the cryostat outside and tilting it on its side so that the remaining LN_2 could flow out of the open top of the annular reservoir onto the ground where it would boil off harmlessly. The student proceeded to do just this.

Unfortunately, the student didn't realize that the upper walls of the LN_2 reservoir, no longer in contact with the nitrogen, had warmed up to near room temperature. When the LN_2 made contact with the warm walls it instantly flashed to vapor and expanded (see Chap. 1). The sudden pressure rise caused the inner wall of the annular space to expand inward towards the center of the cryostat destroying both the cryostat and the experiment it contained.

Fortunately, the student was not injured and in retrospect its seems clear the cryostat was designed to fail in the manner that it did rather exploding outward.

T. J. Peterson and J. G. Weisend II, *Cryogenic Safety*, International Cryogenics Monograph Series, https://doi.org/10.1007/978-3-030-16508-6_2

This accident illustrates that even seemingly straightforward activities in cryogenics facilities can result in serious hazards. The root causes of this particular accident were a lack of experience, a lack of a proper hazard analysis and a motivation to finish a task quickly.

2.2 Introduction

This chapter will discuss the general hazards of cryogenics and their mitigations. Later chapters will discuss in more detail some specific hazards such as Pressure Safety (Chap. 3) and Oxygen Deficiency (Chap. 4). These hazards exist in all cryogenic facilities and must be considered along with those described in this chapter.

This chapter will also discuss a way of working safely that involves hazard analysis, proper training and a general Safe Work attitude. These techniques, expanded upon in Chap. 8, are applicable to all work environments not just those involving cryogenics.

2.3 Effects of Cold

The most obvious hazard typically associated with cryogenics is extreme cold. People can be injured by this cold either by direct exposure to low temperature liquids or vapors or by contact with other materials that have themselves been cooled down by exposure to cryogenic fluids.

This extreme cold can result in eye injury and blindness and tissue damage "burns". Inhaling very cold vapour, in addition to posing a significant oxygen deficiency hazard (Chap. 4) may cause lung damage. Breathing cold vapour or walking through vapour clouds should always be avoided.

2.3.1 Mitigation

The only mitigation to the hazard of extreme cold is to avoid contact with it. This can be accomplished by:

- Designing vents, relief valves or any other components that might suddenly release cold vapor or liquids so that they do not affect walkways or other locations commonly occupied by people.
- Avoiding walking through cold vapor clouds.
- Insulating pipes and vessels containing cryogenic fluids including cold vapor vents with vacuum insulation, foam or other materials so that the surface exposed to people stays at room temperature.

- Being aware of the environment and knowing which surfaces may be very cold due to exposure to cryogenic fluids.
- Wearing appropriate Personal Protective Equipment.

2.4 Personal Protective Equipment

In many cases it is impossible to guarantee that you will not come into contact with extremely low temperatures, even during routine operations, in a cryogenic facility. The solution is to protect oneself from injury by wearing Personal Protective Equipment (PPE).

PPE requirements may vary between institutions but the recommended minimum set of PPE is:

- Eye protection via safety goggles or safety glasses with side shields
- Face protection via face shield (A face shield *does not* provide eye protection, safety glasses still need to be worn under it)
- Easily removable insulated gloves appropriate for cryogenics (NOT clean room gloves, cotton or wool gloves which can wick the cryogenic liquid to your skin and hold it there—increasing damage)
- An insulated apron to protect against splashing
- Long pants without cuffs that cover the tops of your shoes
- Closed toed shoes

This PPE should be worn *whenever* handling open containers of cryogens, transferring cryogens between containers (including filling LN_2 dewars) transferring cryogens from a liquid trailer, and pulling or inserting bayonet connections (or U-tubes).

There are some people that claim the liquid helium will never form liquid droplets in room temperature due its low boiling point and low latent heat and thus PPE is not required to work with liquid helium. While one can argue the droplet issue based on the amount of liquid helium released, liquid helium can certainly produce very cold vapor that can cause damage and thus the PPE requirements for helium should be no different than that for other cryogens.

Operation of sealed cryogenic systems such as those cooled by cryogenic plants may not require this level of PPE. However, if maintenance activities take place that could expose you to cryogenic temperatures then the PPE should be worn.

Perform a hazard analysis (see below) for all tasks to see if additional PPE is required. ALWAYS wearing eye protection in a cryogenics facility is a good practice and required in many institutions.

Note that since even small amounts of cryogens can cause damage, PPE should be used in small laboratory situations and during teaching and public demonstrations using cryogens. Depending on the circumstances, audience members in a demonstration may also require PPE.

Fig. 2.1 Examples of typical PPE for use in cryogenics. *Courtesy* TempShield (http://www.tempshield.com)

Figure 2.1 shows examples of some typical PPE used in cryogenics. Note that in addition to these components safety glasses also need to be worn. Such PPE is commercially available. Reference [1] is a good resource for locating PPE suppliers.

Note that specialized PPE is required when dealing with oxygen and hydrogen systems. See Chaps. 5 and 6 for additional information.

2.5 First Aid

In the event of burns due to contact with cryogenic temperatures, the first aid recommended by the British Cryogenics Council [2] is

"Flush the affected areas of skin with copious quantities of tepid water, but do not apply any form of direct heat, e.g. hot water, room heaters, etc. Move the casualty to a warm place (about 22 °C; 295 K). If medical attention is not immediately available, arrange for the casualty to be transported to hospital without delay. Treatment for shock may be necessary.

While waiting for transport:

(a) Loosen any restrictive clothing.
(b) Continue to flush the affected areas of skin with copious quantities of tepid water.
(c) Protect frozen parts with bulky, dry, sterile dressings. Do not apply too tightly so as to cause restriction of blood circulation.
(d) Keep the patient warm and at rest.
(e) The patient should not be administered pain killers.
(f) The patient can be given oxygen if available.

(g) Ensure ambulance crew or hospital is advised of details of accident and first aid treatment already administered.
(h) Smoking and alcoholic beverages reduce the blood supply to the affected part and should be avoided."

2.6 Handling of Cryostats and Dewars

A contributing factor to the example accident above was poor handling of cryostat. Mishaps involving the movement of cryostats and dewars are relatively common and can result in injury or significant damage. Tipping or dropping a cryostat containing a cryogen can result in sudden release of the fluid or damage to the cryostat. In order to reduce heat leak [3], the necks of cryostats and dewars are typically quite weak. Damage to this area can result in breaking the neck off, releasing cryogenic fluid or in breaking the vacuum between the outside wall and inner part of the cryostat. Breaking the vacuum will allow air to flow into the space, condensing on the cold inner surface, resulting in a sudden large boil-off of the cryogen and sudden pressure rise (See Chap. 3). Additionally, damage to the upper part of the cryostat may render pressure relief valves inoperable resulting in a dangerous pressure buildup. Recall as in the accident above that even well insulated cryostats may have upper sections significantly warmer than those in contact with the cryogen.

The solution is to always use correct handling techniques:

- Avoid moving all but the smallest of cryostats by hand.
- Use properly designed transport and lifting fixtures.
- If using a crane, use trained and experienced personnel to install the lifting fixtures or straps and to operate the crane.
- Don't tip cryostats and dewars significantly off a vertical orientation, particularly if they contain liquid.
- Properly secure cryostats and dewars to carts, trucks or trailers when moving them. See Fig. 2.2.

Above all, think the movement problem through carefully.

Figure 2.2 shows a LN_2 dewar and its transport cart used at the SLAC National Accelerator Laboratory. In this case the base of the dewar has been welded to the cart to ensure that the dewar both stays vertically oriented and does not slip off the cart during transport. Other solutions involving clamps or straps to properly retain a dewar during transport are also possible.

One additional caution. Keep in mind that liquid helium is much lighter (Chap. 1) than other cryogenic liquids such as liquid argon or liquid nitrogen. Thus, a dewar or cryostat designed to store liquid helium may not be strong enough to support the weight of the heavier cryogenic liquids. Only use dewars or cryostats for the liquids for which they were designed.

Fig. 2.2 LN_2 Dewar and transport cart. The close up shows the weld fixing the bottom of the Dewar to the cart. *Courtesy* SLAC National Accelerator Laboratory

2.7 Pressurized Systems

In addition to very low temperatures, high pressures are a common hazard in cryogenic facilities. Some of these hazards may be obvious, for example, the presence of high pressure gas cylinders or the high pressures associated with cryogenic refrigeration compressors; while others may be not so obvious such as the increase in pressure in a closed system due to the boil off of cryogenic fluids or the sudden boiling and pressure rise when a warm component is placed into a cryogenic fluid. Chapter 3 will discuss in detail pressure safety in cryogenic systems including accident scenarios and pressure relief systems; here we will discuss pressure safety in cryogenic operations.

No matter how well designed the insulation system, some heat will always leak into containers of cryogenic fluid causing boil off of the fluid and a subsequent pressure rise if the container is closed. Since relief valves are typically set to open somewhat above atmospheric pressure, even properly designed systems will exhibit some pressure build up. This pressure when applied against flanges, bayonets or other removable components can result in forces that can do damage, if say a flange is removed while under pressure.

Additionally, pressures can arise in room temperature piping or systems due to action of compressors, leakage through a valve from high pressure source or a pumping and purging operation.

Due to these hazards, one should always verify that systems are depressurized prior to:

- Pulling U-tubes or bayonets
- Opening valves, flanges or container lids
- Connecting or disconnecting hoses or piping

Given that all closed volumes of cryogenic fluid will see a pressure increase due to the inevitable boil off, *Never* place cryogenic fluid or cold gases into any closed container not specifically designed for them (household Thermos bottles are particularly dangerous).

Insertion of warm line or U-tubes into cryogens (e.g. LN_2) will cause rapid boiling, pressure rise and/or venting. Always allow for this hazard in the operation.

High pressure gas cylinders should always be properly restrained. Additional information specific to the safe handling of high pressure gas cylinders may be found in reference [4].

2.8 Presence of Magnetic Fields

A significant application of cryogenics is the cooling of superconducting magnets. Such magnets, depending on their design, may generate strong magnetic fields in the facility. This can occur not only in research and industrial facilities but also in medical facilities such as Magnetic Resonance Imaging (MRI) suites.

Hazards associated with magnetic fields include: disruption of or damage to pacemakers and other implanted medical devices and prosthetics and unplanned movement of metallic components such as carts, pressure cylinders and tools. It is easy to forget the common items that we use or carry with us which may be attracted by a magnetic field. A quick internet search can provide photos and videos of items pulled into magnetic resonance imaging (MRI) magnets or research magnets such as office chairs, survey equipment, and tools. The force and velocity of items drawn into a powerful magnetic field can and have resulted in severe injury. Strict care and vigilance regarding material (iron, steel) even in small quantities is required when entering a region of high magnetic fields.

The best mitigation for this hazard is to survey the stray fields around magnets and indicate the field boundaries and post warning signs. Precautions should then be taken to keep people and equipment out of areas where the fields may cause problems. People with pacemakers or other medical devices should not enter areas where the stray field is above 5 gauss (0.5 mT). Careful thought should be given to taking metallic components into area of stray magnetic fields. Magnetic shielding may used to reduce exposure to stray fields.

An example of good magnetic field safety procedure may be found in reference [5].

2.9 Housekeeping

Proper housekeeping in cryogenic facilities is necessary for safe operations and is touched upon in a number of the following chapters. Attention to housekeeping not only reduces the chance of accidents but also is part of establishing a safe work culture. Housekeeping issues include:

- Removal of debris and reduction of flammable materials.
- Keeping emergency exits and escape routes clear and ensuring that they are well marked.
- The presence of emergency lighting so that exiting the facility in a power failure can be accomplished rapidly and safely. Particular attention here should be paid to stairs and work platforms.
- Ensuring that pressure relief systems (Chap. 3) and vent lines are kept clear of ice, dirt, bird's nests etc.
- Removal of trip hazards and clean-up of spills.
- Ensuring that alarm lights and horns can be seen and heard throughout the facility.
- Maintaining proper clearance for electrical panels and shut off valves.
- Keeping emergency equipment such as sprinkler systems, fire extinguishers and first aid kits in good repair.
- Conducting proper maintenance and testing on Pressure relief systems (Chap. 3), Oxygen Deficiency detection systems (Chap. 4), Hydrogen detectors (Chap. 6) and fire and smoke detectors.

2.10 Job Hazard Analysis, Procedures, Training and Safe Work Culture

In addition to the technical hazards and mitigations described in this book, there are a number of approaches to increase safety in cryogenic facilities. These approaches are equally valuable in all industries and even in many personal activities.

2.10.1 Job Hazard Analysis

A job hazard analysis (sometimes also known as a safety hazard analysis) is a simple and effective way to increase work safety. It is a simple approach for planning work that identifies and mitigates possible hazards.

In a job hazard analysis, all the hazards associated with the specific work to be done are listed and then the appropriate mitigations for each hazard is given. The mitigations may involve technical solutions, PPE, training etc. Job hazard analysis may be applied to work ranging from the very simple to the very complex. Depending on the complexity of the work or institutional requirements, the work may be divided

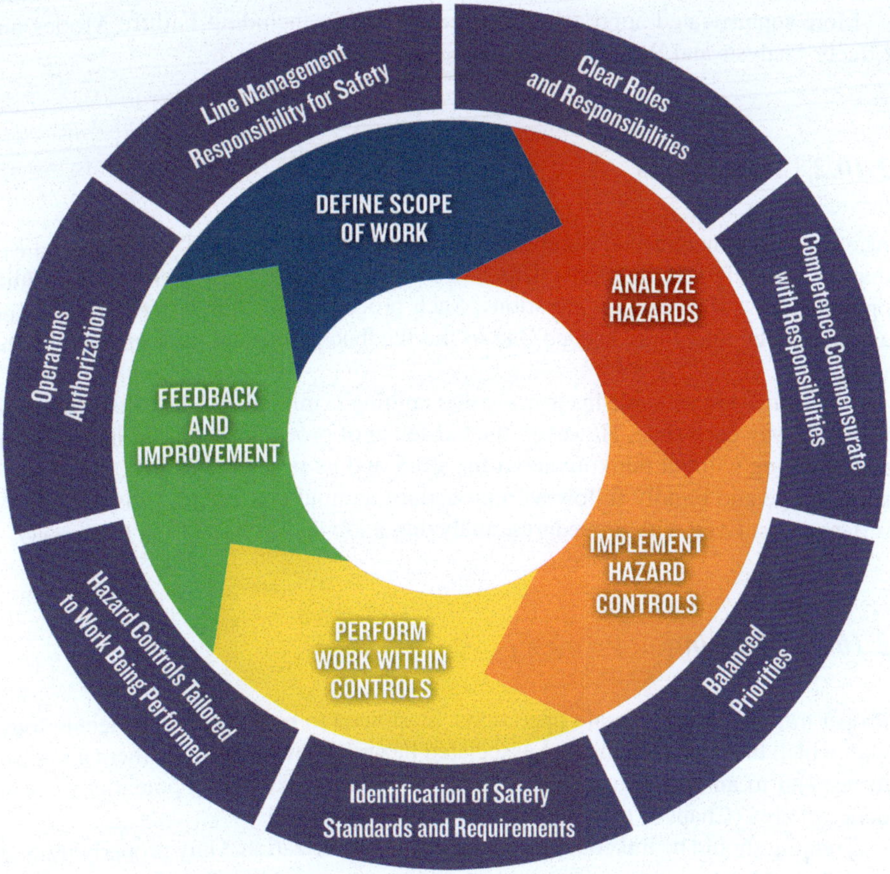

Fig. 2.3 An example of work planning as part of an Integrated Safety Management Program Courtesy National High Magnetic Field Laboratory, Florida State University

into separate steps, with each step having its own set of hazards and mitigations. An example job hazard analysis from the European Spallation Source is given in the appendix of this chapter.

Job hazard analysis is part of a circular approach to work planning that consists of:

1. Plan the work
2. Determine the hazards
3. Apply mitigations to reduce the hazards
4. Carry out the work using the mitigations
5. Feedback the experience gained during the work to future jobs

This approach to planning is shown in Fig. 2.3. More details about job hazard analysis may be found in references [6, 7].

More sophisticated approaches to hazard analysis including Failure Modes and Effects Analysis and What if Analysis are discussed in Chap. 8.

2.10.2 Procedures

Another way to increase safe operations is to develop detailed step by step procedures that include safety mitigations. This is frequently done for particularly hazardous operations or for repetitive operations. Such procedures can be carefully reviewed and should be altered as needed due to the feedback from the performance of the work.

Procedures are valuable in ensuring that nothing is missed and that the job is done the same way each time. However, the existence of procedures should not result in staff working without thinking about the work and its possible hazards. Conditions may change and blindly following a procedure without considering this possibility is dangerous. Even with procedures, analyzing a job and its hazards is necessary.

2.10.3 Training

Proper training is vital to safe operations. Staff need to be trained in the technology with which they are working, the associated hazards, how to mitigate them and how to respond to an emergency. For example, all staff working in a potential oxygen deficient area (Chap. 4) must be made aware of the hazard.

Training should be linked to access and work authorization. Only properly trained staff should be permitted to enter potentially hazardous areas or work on hazardous systems. Maintaining accurate training records is crucial. In many facilities, door access cards are directly linked to the training database.

When thinking about training don't forget occasional visitors such as janitors, security staff and delivery personnel.

2.10.4 Safe Work Culture

One of the best ways to improve safety, including cryogenic safety, in an institution is to develop a safe work culture. This is a belief that nothing, including schedule, cost, profits, scientific output or efficiency, comes before safety. While there is no way to remove all risk and nothing can be 100% safe; reasonable, agreed upon or legally mandated safety policies must be put ahead of all other considerations.

Establishing such a culture can be difficult. Even if appropriate policies are in place, staff may be tempted to work in an unsafe manner to meet schedule or save on costs. The role of senior management is vital here. Management must be very explicit that safety is the top priority and they must show by actions that this is true. If management says one thing but does another concerning safety priorities, a safe work culture can't be established. A clear vision is important here. For example, Skanska, the construction firm leading the civil construction of the European Spallation Source has a philosophy that "We work safe or not at all". This philosophy is appropriately supported by rules, inspections and responses to unsafe behavior.

All employees have a role in creating and maintaining a safe work culture. Employees and contractors should all be empowered to stop an activity (even one outside their immediate area of responsibility) if they deem it to be unsafe. Such empowerment not only reduces accidents but reaffirms the priority of safety in the institution.

2.11 Best Practices

- Always wear appropriate personal protection equipment, including eye protection, when handling cryogenic fluids no matter how small the amount.
- Ensure that relief valves and vent lines do not direct the flow of cold gas towards people or towards materials not designed for cryogenic temperature use.
- Avoid breathing cold vapor and walking through vapor clouds.
- Ensure that cryogenic systems are depressurized before carrying out activities such as: pulling of bayonets and U-tubes, opening valves, flanges or container lids and connecting or disconnecting piping or hoses.
- Only use approved containers for storage and transport of cryogenic liquids and cold gases.
- Only use dewars and cryostats with the specific cryogenic liquid for which they have been designed.
- Never defeat or bypass pressure relief systems.
- Do not become complacent when working with LN_2 Even small amounts can cause eye or tissue damage.
- Don't play with LN_2. Take care if releasing LN_2 on to the ground or otherwise into the environment.
- Take into account the hazards associated with stray magnetic fields.
- Properly plan work including identifying and mitigating hazards via a Job Hazard Analysis or similar approach.
- Maintain proper housekeeping.
- Strive to develop a Safe Work culture.

Appendix

Example of a Job Hazard Analysis From ESS

EUROPEAN SPALLATION SOURCE

Job Hazard Analysis

Note: The job hazard analysis must be regularly reviewed and updated when necessary as indicated in ESS-0064035. A copy of this JHA must be kept in the location where the job is performed.

Job: LEBT vacuum system assembly and test

Number of people performing the job: 3

Date and duration: January 2018 to February 2018

Location/Work Area: Accelerator Tunnel (G01)

Approvers	Name	Date	Signature	Phone No.
Area supervisor:	[redacted]	04/09/17		[redacted]
System owner:	[redacted]	04/09/17		[redacted]
If work is considered to be hazardous or new ES&H Div or designated Safety Committee Rep:	[redacted]	04/09/17		[redacted]

PERSONAL PROTECTIVE EQUIPMENT (PPE) REQUIRED FOR WORK AREA (SEE AHA FOR THIS WORK AREA)

- A Safety gloves (EN 388, EN 420)
- B Safety helmet (EN 812, EN 397, EN 14052 or EN 12492)
- C Safety shoes (EN 345, EN 346 or EN 347)
- D Ear protection (EN 352)
- E Face protection (EN 166)
- F Eye protection (EN 166)
- G Safety overall
- H Respiratory protection (EN 149)
- I Other, please specify: (specify PPE and associated standard)

Work Tasks/Steps Identify work steps/tasks in sequence when such sequencing contributes to safety, security, and/or environmental protection.	**Hazards, Concerns, and Potential Accidents/Incidents** Identify activity hazards for each step. Consider interactions with hazards identified in AHA for this work area.	**Controls, Preventive Measures, and Actions** Specify controls for each hazard [e.g., LOTO points, additional PPE beyond that already required by AHA, Tamper Indicating Devices, alarms, safes, recycle, waste minimization]; reference documents such as permits, operating manuals, or procedures; and training/qualification requirements.
Mounting of primary pump	Manual handling/ pinch points	
Mounting of pumping manifold and connection to the primary pumps	Manual handling/ pinch points	clean gloves
Gate valve mounting	Falling object/Manual handling/ pinch points	clean gloves
Assembly of the definitive end part of the LEBT and the commissioning box (Activity to be confirmed)	Falling object/Manual handling	clean gloves
Mounting of turbomolecular pumps on the LEBT and on the commissioning box	Falling object/Manual handling/ pinch points	clean gloves
Cleaning of parts with IPA/Acetone	Chemical hazard/waste generation	Clean gloves, waste management

JHA Page 1 of 2

EUROPEAN SPALLATION SOURCE

Job Hazard Analysis

Connection of utilities (power for gauges and pumps and compressed air for the valves)	Electrical hazard	Lock-out/tag-out of electricity
Pump down	Temperature, noise, electricity	Ear plugs mandatory above 80 dB, or if feeling uncomfortable
Helium leak check	Inert gas, pressure, obstruction	Portable Oxygen deficiency devices
Venting	Inert gas (Nitrogen), pressure, obstruction	Portable Oxygen deficiency devices
Racks assembly	Electrical hazard	Lock-out/tag-out of electricity
Racks assembly	Falling object/Manual handling/sharp material handling	
BI instrumentation mounting	Falling object/Manual handling	clean gloves

EUROPEAN SPALLATION SOURCE

Job Hazard Analysis

Pre-Job Briefing Content

- Identify critical steps or phases of this activity.
- Discuss required controls and preventive measures that are needed. Are they adequate?
- Is other information needed to perform this activity in a safe, secure, and environmentally responsible manner?
- Does everyone agree to the work tasks/steps, hazards, and controls and commit to follow them?

PRE-JOB BRIEFING ATTENDANCE ROSTER/PEOPLE AUTHORIZED TO PERFORM THIS JOB

By signing below **as required**, I agree to the following:

- I agree to follow the work steps and implement the controls as written as applicable to my work assignments.
- I agree to stop work when conditions or hazards change or when I encounter unexpected conditions during the execution of work, or when work cannot be performed as written, or instructions become unclear during execution.
- I confirm that I am authorized, qualified, and fit to perform the work.

Name	Date	Signature

References

1. *The CSA Buyers Guide*, Cryogenic Society of America (2018), https://csabg.org
2. *Cryogenics Safety Manual*, British Cryogenics Council (2018)
3. J.G. Weisend II (ed.), *Cryostat Design* (Springer, 2016)
4. *Standard for Safe Handling of Compressed Gases in Containers*, P-1, Compressed Gas Associated (2015)
5. *SP-19 Magnetic Field Safety*, National High Magnetic Field Laboratory (2007), https://nationalmaglab.org/images/user_resources/searchable_docs/safety/magneticfield.pdf
6. *Job Hazard Analysis* OSHA 3071, US Occupational Safety & Health Administration (2002)
7. "Work Planning and Analysis", FESHM 2060, Fermilab (2014), http://esh-docdb.fnal.gov/cgi-bin/RetrieveFile?docid=525&filename=FESHM%202060.pdf&version=11

Chapter 3
Pressure Safety in Cryogenics

Abstract Sudden rises of pressure are a common hazard in cryogenic systems. This chapter describes such hazards and the methods by which they may be avoided. Pressure sources, the design of pressure relief systems, the proper calculation of relief valve and vent sizes and descriptions of relief valves are all covered. The impact of pressure safety issues on cryogenic system design is discussed. Example calculations, relevant data and references to pressure vessel codes are included.

3.1 Example Accident

A portable liquid nitrogen dewar exploded in a state university chemistry building laboratory in Texas in 2006, causing substantial building damage [1]. Photos from the incident report are reproduced here in Figs. 3.1, 3.2 and 3.3. The relief valve and rupture disk for the liquid nitrogen dewar, which was originally constructed and tested in December 1980, had been replaced with brass plus plugs.

From the incident report:

"The State Fire Marshal's Office, in cooperation with the university's environmental health & safety office, conducted an investigation that included an assessment of the building damage and reconstruction of the events leading to the explosion. The resulting examination revealed catastrophic failure of the cylinder. The failure permitted rapid expansion of the nitrogen gas, blowing out the bottom of the tank and propelling the cylinder upwards.

The examination revealed that the cylinder's pressure release valve and rupture disc had been replaced by two brass plugs. Without these two features in place, the cylinder's rupture-prevention function became compromised. During the investigation, lab students related that the bottom portion of the cylinder had been frosting for approximately twelve to eighteen months, suggesting to them that the cylinder was 'leaking'. It is speculated that the tank was relieving normal excessive pressure through an old leaking gasket on the top of the tank (the actual pressure-relief function had been plugged). Approximately twelve hours prior to the explosion, one of the students replaced the leaking gasket and refilled the cylinder. As the old gasket that helped relieve internal pressure had been replaced, the now full cylinder was

T. J. Peterson and J. G. Weisend II, *Cryogenic Safety*, International Cryogenics Monograph Series, https://doi.org/10.1007/978-3-030-16508-6_3

Fig. 3.1 The laboratory room after the liquid nitrogen dewar explosion

completely sealed. The cylinder ruptured when its internal pressure rose above 1,000 psi."

This incident illustrates the large energy that can be contained and released as liquid evaporates and then the gas warms and pressurizes in a closed volume, no chemical reaction. The small and rather inconspicuous relief devices on these commercially available dewars provide vital protection!

3.2 Cryogenic Pressure Safety—Introduction and Safety Requirements

Cryogenic systems may include pressure vessels and pressure piping at various temperatures, from somewhat above ambient down to the lowest system temperatures. Vessels and piping immediately downstream of compressors, such as bulk oil removal tanks for oil flooded screw compressors, typically operate near compressor discharge temperatures which are elevated due to the heat of compression. Standard pressure safety practices such as ASME code stamping [2, 3] and following the ASME

Fig. 3.2 The explosion blew the laboratory room wall out into the hallway

pressure piping code [4] apply to such systems. In this chapter, we focus on low temperature vessels and some of the unique problems and requirements that result at cryogenic temperatures.

Engineering for pressure safety in cryogenics, as for any pressure system, includes three fundamental aspects:

- Design the vessels (pressure vessels and vacuum vessels) and piping for the desired Maximum Allowable Working Pressure (MAWP).
- Recognize and consider all possible sources of pressure in the design.
- Design venting systems (flow path to the relief device, relief device size, and any downstream ducting) to protect the vessels and piping against overpressure.

For vessel and piping design, local laws and regulations apply and may contain quite specific requirements. These may include recommendations or requirements to follow ASME pressure vessel code [2, 3], ASME pressure piping code [4], or other national standards [5]. Department of Energy (DOE) contractors are subject to a law designated in the Code of Federal Regulations as 10 CFR Part 851, "Worker Safety and Health Program" [6], referred to here as 10 CFR 851, which in brief says that when national standards cannot be applied, methods ensuring an equivalent level of safety must be followed. DOE national laboratories provide guidance in accordance with 10 CFR 851 such as SLAC's ES&H Manual Chapter 14, "Pressure Safety" [7]

Fig. 3.3 The dewar which exploded on the right; a similar but undamaged dewar on the left

and Fermilab's ES&H Manual (FESHM) pressure vessel standards [8] which are publicly available.

For safety and for documenting compliance to safety rules, it is best to purchase vessels built to a consensus code from code-authorized shops. When code stamping is not possible, one may design (or specify designs) in accordance with the intent of the code and note implications of exceptions to the code. Examples of special cases include helium dewars with fiberglass-epoxy necks for very low loss rates and superconducting radio frequency (SRF) cavities made from niobium and titanium which operate at liquid helium temperatures. 10 CFR 851 provides guidance for such situations. In general, protection of vessels and piping against overpressure depends on system details and is the responsibility of the owner/operator.

3.3 Sources of Pressure

Sources of pressure include mechanical devices such as compressors and pumps, connection to a higher pressure source such as a tube trailer, and pressurization of cold gas or liquid with the addition of heat. For positive displacement devices, worst case flow may occur with high suction pressure as limited by inlet-side reliefs or

Table 3.1 Practical radiative heat loads through evacuated multilayer insulation (vacuum < 0.01 Pa) [24]. Although lower heat flux is often reported in tests of MLI, the author's experience is that these numbers represent typical values attained in practice. Note that heat influx via support structures and "end effects" will be additional to this radiative heat load and may dominate the total heat load

Situation	Radiative heat load
From 300 K to MLI-insulated (typically about 30 layers) cold ("cold" meaning a temperature in the range 80–2 K) surface	~1.5 W/m^2
From 80 K to MLI-insulated (typically about 10 layers) 4.5 K or 2 K surface	~50 mW/m^2

pump/compressor motor power. One may then calculate worst-case flow as highest inlet density combined with known displacement volume. However, power limitations of the pump or compressor motor may reduce the worst-case flow below the estimate from displacement volume. Screw compressors, commonly used for compressed air systems and large helium refrigeration systems, are positive displacement devices, so throughput is proportional to inlet density within the limits of compressor power and volumetric efficiency.

Thermal pressure generation may occur in a cryogenic system due to the normal heat flux into the system or with special circumstances such as fire or loss of vacuum. Trapped volumes will experience a slow warm-up and pressurization with normal heat influx. One must consider all possible volumes which may contain "trapped" (closed off by valves or by other means) cold fluid which may warm up. Such warm-up may be slow or involve only a small volume and so require a small relief. A rate of warm-up may be evaluated but is generally slow enough that trapped volume reliefs are not individually analyzed, rather evaluated based on a general worst case. Typical heat input through insulating vacuum and MLI is shown in Table 3.1.

Loss of vacuum to helium with convection and conduction through helium gas provides a significant heat flux and warm-up rate. The worst case for containers at liquid helium temperatures is generally sudden large heat flux due to condensation of nitrogen or air on the liquid helium temperature surface (loss of insulating vacuum or some internal vacuum like for a particle beam to air or nitrogen). The surface may include multi-layer insulation (MLI), in which case heat must pass through MLI. Air condensation in the absence of MLI on a bare metal surface provides the greatest heat flux.

Heat flux to liquid helium temperature containers via air condensation on the other side of the container surface from the helium is so dynamic that several groups have performed tests to measure the heat flux. Table 3.2 provides a summary of some of those results. We see typical peak values in the range of 3.5–4.0 W/cm^2.

In another study, T. Boeckmann et al. [9] saw that air inflow into a helium-cooled niobium radio frequency (RF) cavity beam vacuum was greatly damped by RF cavity structures. For liquid helium temperature containers with very large surface area, atmospheric air rushing into a vacuum space and condensing on a surface deposits about 12 kW per cm^2 of air hole inlet area, based on air flow rate in and heat of condensation. In many cases, heat flux will be limited by this air hole inlet size

Table 3.2. Heat flux to liquid helium temperature container with loss of vacuum to air or nitrogen

References	Heat flux to nominally 4.2 K helium
Lehman and Zahn [25]	0.6 W/cm^2 for the superinsulated tank of a bath cryostat 3.8 W/cm^2 for an uninsulated tank of a bath cryostat
Cavallari et al. [26]	4 W/cm^2 maximum specific heat load with loss of vacuum to air
Wiseman et al. [10]	3.5 W/cm^2 maximum peak heat flux 2.0 W/cm^2 maximum sustained heat flux

rather than the total low-temperature surface area. Authors also comment about a layer of ice quickly reducing heat flux [10]. Heat flux curves for liquid helium film boiling [11] with a delta-T in the range of about 50–60 K, which one would expect with air condensation, agree with these heat flux numbers of up to 4 W/cm^2. For a conservatively large heat flux, we suggest assuming 4 W/cm^2 for bare metal surfaces.

Stored energy of a magnetic field (superconducting magnet "quench") may provide an even larger heat deposition to low temperature and larger flow rate than loss of insulating vacuum. The rate of heat deposition with quench depends on magnet structure and heat transport to the helium. For example, one may observe a bimodal generation of helium flow with cold iron accelerator magnet structures: an initial quick burst of helium which was in intimate contact with the magnet coil, followed some 10's of seconds later by a second, longer burst of helium from the iron, steel, and/or aluminum structure around the coil.

If a source of fuel for fire is nearby, such as grass or wood near a pressure vessel, one must consider pressure generated by heat from fire, with heat transport through a gas-filled insulation space for generation of internal tank pressure.

3.4 Analytical Methods for Vent Line and Relief Sizing

Vessels and piping have a Maximum Allowable Working Pressure (MAWP) defined by the design of the vessels or system. A venting system and relief devices must be in place to prevent any event from pressurizing the vessel or piping above the MAWP (plus whatever code allowance may be available). The procedure in general is to:

- Evaluate all pressure sources and possible mass flow rates
- Evaluate the mass flow required to be vented to prevent a pressure above the MAWP (plus code allowance)
- Size the vent line to the relief device
- Size the relief device
- Size downstream ducting, if any (downstream piping may be necessary to carry inert gas safely away from an occupied area or sensitive equipment)

All of the above analytical steps include evaluation of temperature and pressure of flow stream (typically a pressure drop analysis for turbulent subsonic flow) as part of the pressure drop analysis. In such a set of analyses, we want to be conservative, to err on the safe side. Venting is typically not steady-state but rather very dynamic, with rapidly increasing flows, and rapid pressure and temperature changes.

We also want the analyses to be reviewable. Managers, safety authorities, and others will want to understand the assumptions and be convinced of the validity of results. Thus, one should present the simplest and most straightforward analysis which demonstrates the safety venting requirement. Of course, more sophisticated analysis, such as computational fluid dynamic (CFD) simulation, may be necessary for a system with severe constraints.

Considerations for emergency venting include various failure modes such as:

- Loss of insulating vacuum
- Loss of beam tube vacuum, RF cavity vacuum, or other device vacuum
- Helium expansion due to inflow of warm helium gas
- Helium expansion due to heat from load balancing heaters
- Deadheaded flow from compressors

Methodology for emergency vent sizing is provided in various standards:

- ASME, "Section VIII—Division I—Rules for Construction of Pressure Vessels" [2]
- American Petroleum Institute, 520 Part I. Sizing, Selection, and Installation of Pressure-Relieving Devices, Ninth ed., American Petroleum Institute [12]
- American Petroleum Institute, 521 Pressure-relieving and Depressuring Systems, Sixth ed., American Petroleum Institute [13]
- Compressed Gas Association, Inc., S-1.3 Pressure Relief Device Standards Part 3—Stationary Storage Containers for Compressed Gases, Eighth ed., Chantilly, VA: CGA [14]

For pressure drop analysis of isothermal turbulent flow of a compressible or incompressible fluid through a piping system upstream or downstream of a relief device, we may start with the following general form of the Bernoulli equation [15]. (Friction factor definition follows that in "Transport Phenomena" [15]).

$$\int_{P_1}^{P_2} \frac{dP}{\rho} + \sum_i \left(\frac{v^2}{2} \frac{L}{R_h} f \right) + \sum_i \left(\frac{v^2}{2} k \right)_i = 0 \tag{3.1}$$

P_1 is pressure in.
P_2 is pressure out.
ρ is fluid density, here only a function of pressure since temperature assumed constant.
v is average fluid velocity within the i-th section of conduit or downstream of the i-th fitting.
L is conduit section length.

R_h is channel hydraulic radius, defined as flow area divided by wetted perimeter, which implies $R_h = D/4$ for round pipes.

f is friction factor based on hydraulic radius.

k is the resistance factor for fittings such as elbows, tees, sudden flow area changes, etc. f and k are dimensionless. These equations contain no unit conversion factors; use any consistent units for other parameters.

For constant density within each section of the conduit and substituting for v from the expression for mass flow,

$$\dot{m} = \rho v \left(\pi \frac{D^2}{4} \right) \tag{3.2}$$

For round conduit, Eq. 3.1 becomes

$$\Delta P = \sum_i \left(0.811 \frac{\dot{m}^2}{\rho D^4} \frac{L}{R_h} f \right)_i + \sum_i \left(0.811 \frac{\dot{m}^2}{\rho D^4} k \right)_i \tag{3.3}$$

The total pressure drop is just the sum of pressure drops through the various straight sections of pipe (the first term, incorporating friction factor, hydraulic radius, and diameter of that section of pipe) and the various fittings (the second term, incorporating the resistance factor k for each fitting and pipe diameter downstream of that fitting). Note that delta-P changed signs here, to a positive number.

Each of the expressions in the sum in Eq. 3.3 is a form of the D'Arcy-Weisbach formula, a practical working formula for pressure drop in a pipe (Eq. 3.4).

$$\Delta P = 0.811 \frac{\dot{m}^2}{\rho D^4} k \tag{3.4}$$

One may view the frictional loss factor in straight pipe in Eq. 3.3, $\frac{L}{R_h} f$, as another resistance factor, k. Conversely, references may express loss factors, k, as equivalent lengths of pipe adding to a length L in the term $\frac{L}{R_h} f$. With pressure drop expressed as head loss, Eq. 3.4 is sometimes called simply the Darcy formula.

For a typical relief venting system, it often works well to calculate pressure drops step-by-step through each section of constant diameter pipe and each fitting (elbow, tee, flow area change, etc.) assuming constant density within each section, using Eq. 3.4 and summing those pressure losses as shown in Eq. 3.3. One may reevaluate density in segments based on the new pressure in each. An example of such an analysis is provided in Section E.

Compressed Gas Association publication, CGA S-1.3, "Pressure Relief Device Standards" [13] includes extensive guidance on requirements for relief devices consistent with ASME code, applicable where MAWP and venting pressure exceed 15 psig. Piping pressure drop equations presented in CGA S-1.3 [14] and other references such as Crane Technical Paper #410 "Flow of Fluids through Valves, Fittings, and Pipes" [16], are typically equivalent forms of the D'Arcy-Weisbach formula with different units and unit conversion factors. For example, one formula for piping pressure drop from the Crane Co. publication [16], is

$$\Delta P = 0.000000280 \frac{K W^2 V}{d^4} \tag{3.5}$$

where ΔP is pressure drop in psi, V is the specific volume (in^3/lb_m), K is the total resistance coefficient (dimensionless) and equal to $\frac{L}{D} f$ for straight pipe, W is the mass flow rate (lb_m/h), and d is the pipe inner diameter (inches). One can see, with specific volume the inverse of density and the various unit conversions, the equivalency of Eqs. 3.5 to 3.4. However, one additional important point is that many sources, including Crane Technical Paper #410 [15] and CGA S-1.3 [14], define friction factor, f, based on diameter rather than on hydraulic radius, a factor 4 difference for pipes! Thus, one needs to be careful to use the friction factor consistent with pressure drop formula for a given reference.

Among the particular issues which must be addressed for low temperature vacuum jacketed helium containers are the temperature at which liquid-to-gas evolution should be estimated for the supercritical fluid at its venting pressure (CGA S-1.3 [14] is very useful here), the warming of the cold fluid passing through a long vent line (CGA S-1.3 also provides useful practical approximation methods here), and the mass generated per unit heat added, or, taking the inverse, a "pseudo latent heat".

Conversion of heat flux to mass flow rate for venting depends on whether the vessel is below the critical pressure, hence contains boiling fluid, or above the critical pressure, containing expanding single-phase fluid. For pressures below the critical pressure, we use the latent heat of vaporization. Net flow out is vapor generated by the addition of heat minus the amount of vapor left behind in the volume of liquid lost. Near the critical pressure, latent heat becomes relatively low, providing a large generation of vapor in terms of mass evaporated, but the vapor is relatively dense, so a significant mass is left behind in space formerly occupied by liquid.

For higher pressures, above the critical pressure, heat added results in fluid expelled. A "pseudo latent heat" can be evaluated to relate heat added to flow rate out. Compressed Gas Association publication CGA S-1.3 [14] provides guidance in assessing venting rates for fluid above the critical pressure, a common situation for helium systems, so we provide a summary here for reference.

From CGA S-1.3, paragraph 6.1.3, for a volume of helium (or another fluid) at or above its critical pressure, a given amount of heat added results in expulsion of the fluid at a rate which is a function of pressure. The heat added per unit mass of fluid expelled from the volume (the "pseudo latent heat") is

$$v \left(\frac{\partial h}{\partial v} \right)_P$$

(with units for example, J/g). Values for pseudo latest heat for helium are tabulated in NBS Technical Note 631, [17] (Table 3.3). The values are also available from the equation of state programs such as HEPAK, by Cryodata, Inc. [18]. Also, "Technology of Liquid Helium," [19] contains a chart of "Heat absorbed per pound of efflux for a helium container relieving above the critical pressure", Fig. 6A-2.

The temperature into the relief device may be higher than the exit temperature from the container due to heat transfer to the flow via the vent pipe. For very high flow rates and a relatively short vent line, this temperature rise may be insignificant. A simple energy balance on the flow and stored energy in the vent line, with an approximate and conservatively large convection coefficient may provide a safely conservative estimate of the temperature rise.

For a long vent line, a more detailed analysis may be required in determining pressure drop and temperature rise to the relief device. CGA S1.3, paragraph 6.1.4, provides some guidance for this analysis, as follows. The temperature of the expelled fluid for analysis of the flow out the vent line is where the quantity

$$\frac{\sqrt{v}}{v\left(\frac{\partial h}{\partial v}\right)_P}$$

is a maximum for the specified venting pressure. This exit temperature will typically be 5–7 K for a liquid helium container venting at a somewhat supercritical pressure (Table 3.4).

Table 3.3 Helium density and heat absorbed over a range of temperatures for 4.0 atmospheres

Temperature (K)	Density (g/L)	$v(dh/dv)_P$ (J/g) (heat absorbed per unit of mass expelled)
5.0	124.2	29.8
5.5	109.4	24.6
6.0	82.6	19.8
6.5	55.0	21.4
7.0	42.7	24.8

Table 3.4. Values of the expressions for determining the approximate pseudo latent heat and discharge temperature from a vessel at a pressure P = 3.0 atm absolute. Discharge temperature is 5.9 K, and pseudo latent heat is 17.9 J/g

Temperature (K)	Specific volume (v) (cm^3/g)	$v\left(\frac{\partial h}{\partial v}\right)_P$ (J/g)	$\frac{\sqrt{v}}{v\left(\frac{\partial h}{\partial v}\right)_P}$
5.70	16.38	16.3	0.248
5.80	18.97	17.0	0.256
5.90	**21.15**	**17.9**	**0.257**
6.00	23.04	18.8	0.255
6.50	30.29	23.0	0.239

3.5 Relief Devices

Relief devices provide protection against overpressure by means of opening to allow escape of the pressurizing fluid or (in the case of protection against excessive vacuum or negative relative pressure) opening to provide a supply of fluid. A conventional relief valve (Fig. 3.4) provides passive venting by means of closure with a mechanical spring or gas spring which permit opening as a function of process pressure. Relief valves may also be "non-reclosing", in other words, they remain open after initial opening, or have other features to modify the effect of pressure on the discharge side of the valve on the valve behavior.

Rupture disks open by means of mechanical breaking of a thin plate or disk, and thus are non-reclosing relief devices. Rupture disks have the advantage of a relatively simple form (Fig. 3.5), so that opening may result in effectively an open port, providing a large flow rate through an otherwise small device.

Section VIII of the ASME Code provides fundamental guidance regarding pressure relief requirements. ASME Section VIII, Division 1, UG-125 through UG-133, provides general selection, installation and valve certification requirements. Since relief valve capacities are typically provided in Standard Cubic Feet per Minute (SCFM) of air, refer to ASME Section VIII, Appendix 11 for flow capacity conversions to SCFM-air. For ASME Section VIII, Division 2, relevant relief sizing information is found in Part 9 (Table 3.5).

Relief valve and rupture disk sizing are best done via valve manufacturer information. The shape of valve body and type of plug make sizing unique to the valve design. The valve sizing formulas provided by manufacturers will typically include factors for required relieving capacity or flow discharge area, gas molecular weight, valve inlet temperature and pressure, compressibility factor, and a coefficient of discharge. (For example, see the Anderson Greenwood, Crosby and Varec products "Pressure Relief Valve Engineering Handbook" [20].) Even though the relief valve or rupture disk normally sits at room temperature, it will cool upon relieving, so one needs to specify cold-tolerant material and design in the procurement. Take care to provide ASME UV-stamped valves for code-stamped vessels.

Manufacturers certify flow capacity for UV-stamped (ASME approved) valves, a major advantage of UV-stamped valves. However, such certification is not generally available for valves discharging at less than one atmosphere gauge pressure. The scope of ASME Section VIII, Division 1, as described in paragraph U−1(c)(2)(−h)(−1), excludes pressure vessels "having an internal or external pressure not exceeding 15 psi". The "15 psi" used to be generally regarded as psi gauge but more recently has been interpreted as psi differential. Since vacuum jacketed vessels have a differential pressure of more than 1 atm with just a slightly positive gauge pressure, one may have a pressure vessel with an MAWP less than one atmosphere gauge. Thus, the requirement of a relief valve for a vacuum jacketed vessel with MAWP less than 1 atm gauge may conflict with the need for a UV-stamped relief valve.

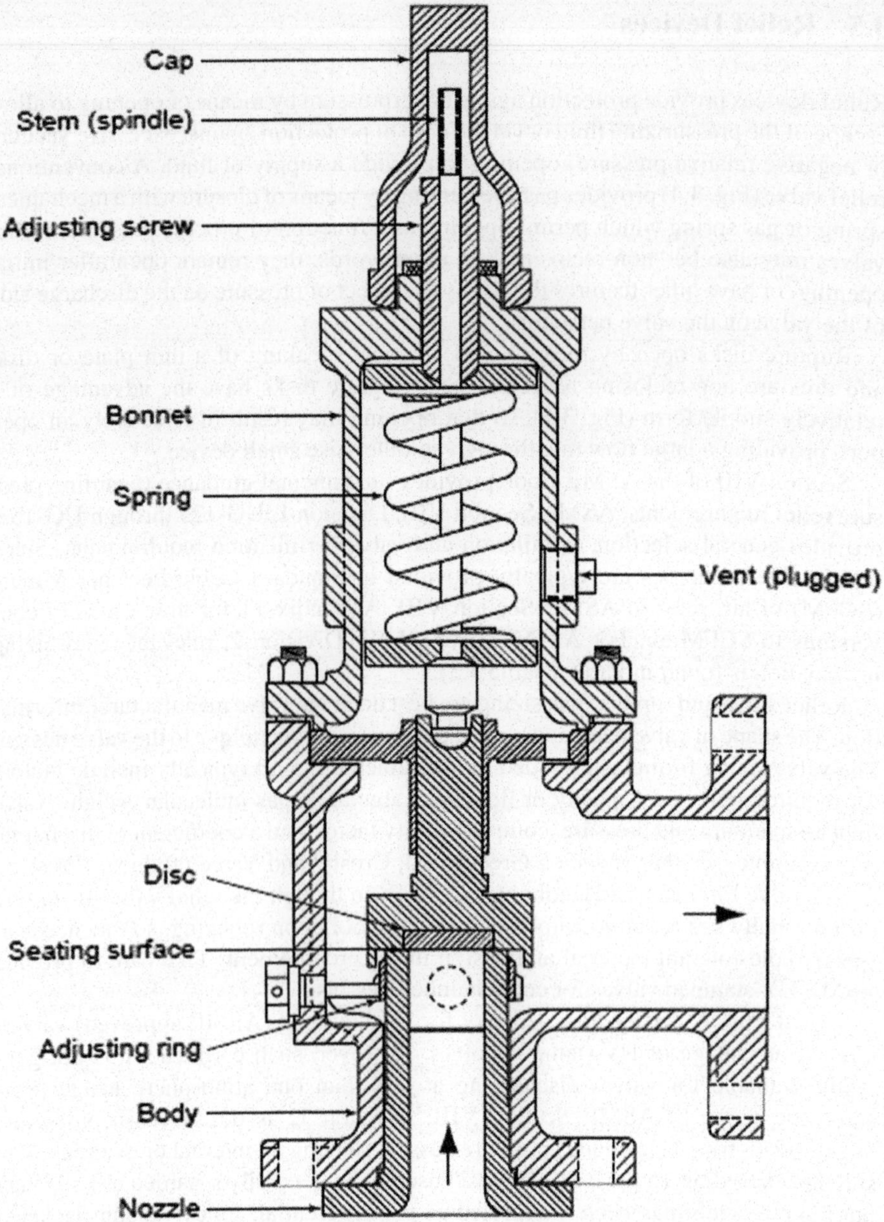

Fig. 3.4 Conventional Safety Relief Valve (From API Standard 520, Part I, Fig. 2). API Standard 520 provides a very nice series of similar illustrations of various types of relief devices which nicely explain their structure and function [12]

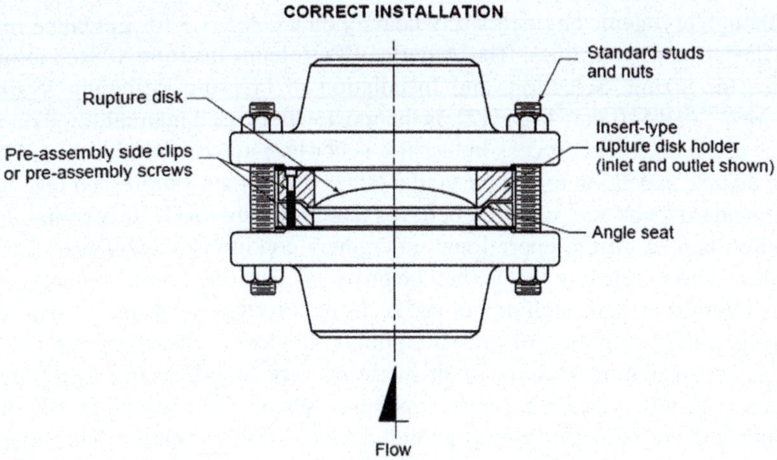

Fig. 3.5 Rupture disk (From API Standard 520, Part I, Fig. 11) [12]

Table 3.5 Vessel pressures and relief set pressures allowed per ASME Section VIII, Division 1

% of MAWP	Maximum vessel pressure as % of MAWP for relief configuration or purpose (UG-125)	Maximum relief set pressure as % of MAWP (UG-134)
100%		Maximum set pressure for single relief device
105%		Maximum set pressure for second and subsequent reliefs
110% (or 3 psi, whichever is greater)	Vessel has single relief device	Relief for "fire" (or unexpected external heat source)
116% (or 4 psi, whichever is greater)	Vessel has multiple relief devices	
121%	Vessel "… exposure to fire or other unexpected sources of external heat …" I include loss of vacuum in this category	

A method which has been used to help ensure safety of low pressure vacuum jacketed containers for which UV-stamped relief valves are not available is to provide both a low pressure "operational" relief, which may be essentially a check valve, and a low pressure rupture disk or lift plate, a device which will not fail in the closed position. If one has enough pressure capacity and low operating pressure requirements such that the operational relief may open far enough below the rupture disk opening pressure, such a system provides a good level of safety.

Although cryogenic engineers rely heavily on CGA S-1.3 for guidance in sizing relief devices and vent lines, The American Petroleum Institute's "Recommended Practice for Sizing, Selection, and Installation of Pressure-Relieving Systems in Refineries", API 520 Part 1 (3) [12], is the most widely used manual for sizing relief devices in the chemical process industries, according to Crowl and Louvar [21].

We do take exception to one particular statement in the Compressed Gas Association Standard CGA S-1.3: "CGA believes that reclosing PRDs on a container shall be able to handle all the operational emergency conditions except fire, for which reclosing or non-reclosing PRDs shall be provided. The operational emergency conditions referred to shall include but not be limited to loss of vacuum, runaway fill, and uncontrolled operation of pressure buildup devices." The exception is that we treat loss of insulating vacuum to air, with the very high heat flux resulting from condensation on the liquid helium temperature surface of a container, like the fire condition and may use non-reclosing relief devices for that situation. This interpretation is consistent with wording in the ASME code, Section VIII, Division I, UG-125, which refers to "… exposure to fire or other unexpected sources of external heat."

Rupture disks inexpensively provide very large capacity, so are typically installed for the worst-case loss of vacuum. Rupture disks are available in various designs, some pre-etched or with knife edge as the failure location, or providing failure in collapse (pressure on the dome). A precise opening pressure is difficult to ensure in a rupture disk, so one should allow some tolerance for failure slightly below the set pressure; one does not want these opening in normal operations. A rupture disk may be viewed as a last resort device since it will not reclose.

Operational reclosing relief valves set at a safely lower pressure (80% of rupture or less) may be installed in parallel with the rupture disk to prevent accidental opening of the rupture disk with small operational disturbances. One may also provide switching valves for dual disks such that one can be replaced while the other holds pressure and provides protection.

3.6 Examples of Venting System Analyses

A superconducting radio frequency (SRF) cavity helium vessel within a vacuum jacketed container provides a nice example of a helium vessel venting analysis.

The vent line assembly shown in Fig. 3.6 sits atop a helium vessel which contains a superconducting niobium RF cavity, vacuum inside the cavity and liquid helium surrounding it. We assume that pressure generation in the liquid helium vessel surrounding the RF cavity comes from loss of cavity vacuum to air with resulting air condensation on the bare niobium surface. We conservatively assume 4.0 W/cm^2 for a heat flux to liquid helium with such an event. We have roughly 8750 cm^2 of cavity surface area which would be exposed to air with loss of cavity vacuum. This implies a heat flux of 35 kW.

This is a low pressure liquid helium vessel, liquid helium surrounding an SRF cavity which is capable of limited external pressure. We want to vent helium with an

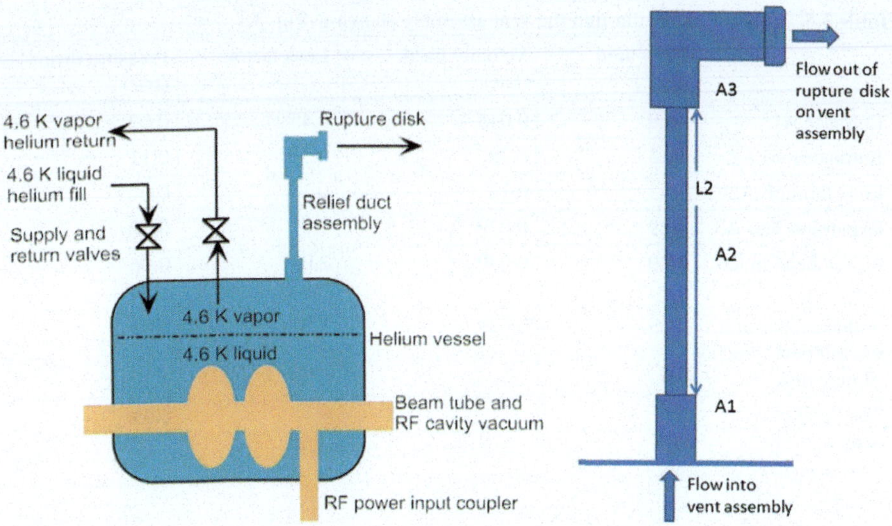

Fig. 3.6 Helium vessel surrounding superconducting RF cavity and helium vent flow path. Flow into the vent assembly is generated by boiling liquid helium from vessel to which this assembly is attached

approximate vessel pressure of 1.5 bar absolute. The latent heat of helium at 1.5 bar absolute pressure is 17.21 J/g. The vapor density is 0.23 of the liquid density, so 0.77 of the helium which absorbs the latent heat is ejected. Thus, the effective latent heat is (17.21 J/g)/(0.77 g-ejected/g) or 22.4 J/gram-ejected. Another way to represent this is 22.4 W/(g/s flow). A similar analysis shows that we have only slightly changing (22–20 W/(g/s)) from low pressure up to the critical pressure of 2.1 bar. We use 22 J/g effective latent heat for this analysis. 35 kW thus produces 1590 g/s flow with only minor dependence on actual pressure in the helium vessel up to 2 bar absolute (approximately 1 atm gauge).

The pressure drop out the helium vessel vent assembly shown in Fig. 3.4 may be approximated as: sudden contraction from a large vessel volume into area A1, second contraction into area A2, flow through the L2 = 11.34-inch length of A2, sudden expansion into A3, a sharp right angle bend in area A3, and exit via a rupture disk. Helium density is conservatively evaluated as the density of saturated vapor at 1.3 bar, 0.022 g/cm^3.

For this very high flow rate and short vent tube, we need not consider warming of the exhaust flow. From Compressed Gas Association standard CGA-S-1.3-2008, "The temperature to use in flow capacity calculations for subcritical fluids shall be the saturation temperature at the flow rating pressure. If the PRD is connected to the container through piping longer than 2 ft (0.6 m) in length, which can transfer heat into the product being relieved or reduce pressure at the relief device inlet, a correction factor, F, shall be used …." We have less than a 2 ft length to the pressure

Table 3.6. Pressure drops through the vent assembly shown in Fig. 3.1

	Flow area (cm^2)	Velocity head (bar)	Loss factor	Pressure drop (bar)
Entrance A1	35.9	0.045	0.45	0.02
Entrance into A2	12.3	0.38	0.30	0.11
Flow through A2	12.3	0.38	0.18	0.07
Expansion into A3	44.9	0.029	7.0	0.20
90 deg bend in tee A3	44.9	0.029	2.0	0.06
Sudden expansion via full port rupture disk	44.9	0.029	1.0	0.03
Total pressure drop				**0.49**

relief device (PRD), so we may assume saturated vapor density for relief flow out of the helium vessel in the worst case, loss of cavity vacuum to air.

For each section of the vent line, pressure drop is estimated using Eq. 3.3. We refer to the term $0.811\frac{\dot{m}^2}{\rho D^4}$ as "velocity head" and provide the result in pressure units (bar). The areas, velocity heads, loss factors, and pressure drop in each of the three sections of the vent assembly are tabulated below. Total pressure drop estimate is 0.49 bar, or 7.1 psid (Table 3.6).

A more complex analysis is summarized below, in Table 3.7. The principle is the same—stepwise analysis of pressure drops through piping segments and fittings. In this second example, density is reevaluated following each segment pressure drop. Segments of piping include those within the cryostat, transitional to the outside of the vacuum vessel, a vent line to a rupture disk, and a larger diameter vent line downstream of the rupture disk.

The downstream vent line carries helium flow out of the experimental area in order to mitigate the oxygen deficiency hazard (ODH) with a helium release. In general, ducting relief valves and rupture disks into an exhaust vent or vents, where practical, provides a good engineering solution to help mitigate ODH risk.

The pressure drop results for the examples in Tables 3.6 and 3.7 are unavoidably large relative to absolute pressure. We have very high flow rates due to the large loss-of-vacuum heat flux combined with low external pressure limits for RF cavities for which radio frequency function determines the material, material thickness, and structure.

A concern for such large pressure drops to relief devices is chattering of the relief and consequent pressure oscillations and surging, which may impose large forces or vibrations on piping as well as reduce the net flow rate. The use of a non-reclosing relief device (a rupture disk) here eliminates the issue of relief valve chattering for worst-case venting flow rates. ASME Section VIII, Division 1, Nonmandatory Appendix M, "Installation and Operation" (of relief devices), in M-6, "Inlet Pressure

Table 3.7. Vent line pressure drop analysis through a burst disk and subsequent piping with stepwise reevaluation of helium density based on pressure. The column "beta" is the ratio of smaller/larger cross sectional areas, used to find the loss factor (K) for the expansion or contraction

Mass flow rate helium	865.73	g/sec											
Vessel's helium pressure	2	bar at helium vessel, start of calc's											
Helium temperature	8	K											
absolute viscosity	2.00E-05	gr/cm-sec assumed constant											
	Fitting ev	L (in)	L (cm)	d (cm)	pressure (bar)	(Adjusted for pressure) Helium density (g/cm^3)	Velocity head (bar)	Reynolds	(Based on Re and on Rh) Friction factor	Beta	K	Pressure drop (bar)	Pressure drop due to friction in straight sections (bar)
Piping within HTC													
Entrance into vent pipe	0.45			5.49	2.000	0.013	0.052				0.45	0.023	
Mitre	1.50			5.49	1.977	0.013	0.053				1.50	0.079	
Straight run		2.00	5.08	5.49	1.898	0.012	0.056	1.01E+07	1.40E-03		0.0052	0.000	0.000
90 deg bend	0.70			5.49	1.898	0.012	0.056				0.90	0.051	
Vent line to LL dewar		42.43	107.77	5.49	1.847	0.011	0.059	1.01E+07	1.40E-03		0.1104	0.007	0.007
Thru LL dewar	1.45			5.49	1.840	0.011	0.059				1.45	0.086	
straight run		3.80	9.65	5.49	1.754	0.010	0.065	1.01E+07	1.40E-03		0.0099	0.001	0.001
tee thru	0.30			5.49	1.753	0.010	0.065				0.30	0.019	
2-90 deg bends	1.40			5.49	1.734	0.010	0.066				1.40	0.092	
straight run		2.40	6.10	5.49	1.642	0.009	0.072	1.01E+07	1.40E-03		0.0062	0.000	0.000
Sudden contraction	0.06			5.08	1.641	0.009	0.072			0.86	0.06	0.005	
Flexhose assy		25.70	65.28	5.08	1.637	0.009	0.099	1.09E+07	1.38E-03		0.0708	0.028	0.028
Sudden expansion	0.03			5.08	1.609	0.009	0.102			0.86	0.03	0.003	
Pumping line weldment													
helicoflex flange		1.20	3.05	5.49	1.606	0.009	0.075	1.01E+07	1.40E-03		0.0031	0.000	0.000
90 deg bend	0.70			5.49	1.606	0.009	0.075				0.7000	0.052	
straight run		6.30	16.00	5.49	1.553	0.008	0.079	1.01E+07	1.40E-03		0.0164	0.001	0.001
tee branch	1.00			5.49	1.552	0.008	0.079				1.0000	0.079	
Helium vent line branch													
straight run		13.10	33.27	5.49	1.473	0.008	0.087	1.01E+07	1.40E-03		0.0341	0.003	0.003
90 deg bend	0.70			5.49	1.470	0.008	0.087				0.70	0.061	
straight run		22.25	56.52	5.49	1.409	0.007	0.094	1.01E+07	1.40E-03		0.0579	0.005	0.005
Expansion 2 to 3-inch	1.63			5.49	1.403	0.007	0.094			0.439008	1.63	0.154	
3-inch diameter piping													
BS&B LPS burst disk	0.16			7.62	1.249	0.006	0.031			0.846852	0.16	0.005	
Straight run		72.00	182.88	8.28	1.244	0.006	0.022	6.66E+06	1.56E-03		0.1376	0.003	0.003
Expansion 3 to 4-inch	0.50			8.28	1.241	0.006	0.022			0.58562	0.50	0.011	
4-inch diameter piping													
Straight run		216.00	548.64	10.82	1.230	0.006	0.008	5.10E+06	1.66E-03		0.3377	0.003	0.003
90 deg elbow	0.70			10.82	1.227	0.006	0.008				0.70	0.005	
90 deg elbow	0.70			10.82	1.222	0.006	0.008				0.70	0.006	
90 deg elbow	0.70			10.82	1.216	0.006	0.008				0.70	0.006	
90 deg elbow	0.70			10.82	1.211	0.006	0.008				0.70	0.006	
Expansion 4 to 6-inch	1.51			10.82	1.205	0.005	0.008			0.449071	1.51	0.012	
6-inch diameter piping													
Straight run		96.00	243.84	16.15	1.193	0.005	0.002	3.42E+06	1.84E-03		0.1111	0.000	0.000
Exit	1.00			16.15	1.193	0.005	0.002				1.00	0.002	
Total	15.89368				1.19							0.809	0.052
													6.39%

Drop for High Lift, Top Guided Safety, Safety Relief, and Pilot Operated Pressure Relief Valves in Compressible Fluid Service," recommends that "the flow characteristics of the upstream system shall be such that losses shall not exceed 3% of the valve set pressure." Such a low pressure drop may be difficult to achieve, in which case non-reclosing devices would be recommended as we have seen in these examples.

3.7 Examples of the Impact on Cryogenic Design

Design impacts due to venting requirements may include features for limiting air inflow, insulation on helium temperature containers when steady-state benefits are small, and internal pipe or channel sizes for dynamic conditions as well as steady-state.

Worst-case air inflow might result from accidental removal of a vacuum flange on beam vacuum or insulating vacuum. Preventive measures for such an event have included a guard rail in accelerator tunnel so that service vehicles cannot strike devices connected to a vacuum flange and knock off the flange, and designs incorporating a few smaller flanges (such as for vacuum relief venting) spaced apart rather than one larger flange.

Five to ten layers of multi-layer insulation (MLI) on a liquid helium vessel will slow the rate of air condensation and consequent heat flux and venting flow rate. When this helium vessel is surrounded by a low temperature thermal shield, the steady-state heat reduction benefit from MLI may be negligible, so the use of MLI will be motivated by the loss of vacuum situation. For this purpose, materials other than MLI which provide a low temperature compatible layer of insulation may be used.

One must also design internal lines and other features to accommodate the large mass flow with emergency venting. Pipes within a cryostat which carry helium in normal operation may also provide the flow path to external relief devices, in which case emergency venting may influence line size.

An example of design driven in part by emergency venting requirements (Fig. 3.7) involved placement of a magnetic shield around a niobium RF cavity inside the helium vessel. Holes were machined in the magnetic shield material such that helium boiling from the niobium surface with loss of cavity vacuum could pass through, but small enough holes that the magnetic shield properties were not impaired. The magnetic shield fits rather closely within the helium vessel, so consideration was also given to the clearance around the magnetic shield for helium flow.

Superconducting magnets provide many examples of design to accommodate venting flow requirements. Focusing magnets and bending magnets for particle accelerators, where very high magnetic fields are required, generally consist of magnet coils fitting closely around the particle beam vacuum pipe. Figure 3.8 illustrates a cross section of the superconducting quadrupole magnets provided by Fermilab to CERN for the Large Hadron Collider [22]. This compact arrangement of coils around the beam tube provides maximal magnetic fields at the particle beam location for focusing the particle beams into the LHC detectors.

When the superconducting magnet coils suddenly become resistive, called a "quench", the stored magnetic energy is dissipated in the now resistive (no longer superconducting) magnet coils. Rapid heating results in rapid pressurization of the helium within and around the coils. The compact nature of the magnet, dictated by magnetic field requirements, limits the flow paths available for venting helium during a quench. Compromises in magnetic design were made in order to provide channels for heat transport and quench flow venting.

A paper published early in the design of the LHC superconducting dipole magnets [23] describes an analysis of similar magnet channel dimensions for venting. This paper studies dipole magnets, different from the quadrupole shown in Fig. 3.7 but with an analogous design. Quoting from their conclusion, "The pressure peak shows a strong dependence on magnet length and radial gap between the beam tube and the winding." This citation illustrates the early consideration of helium venting requirements as part of a complex design project.

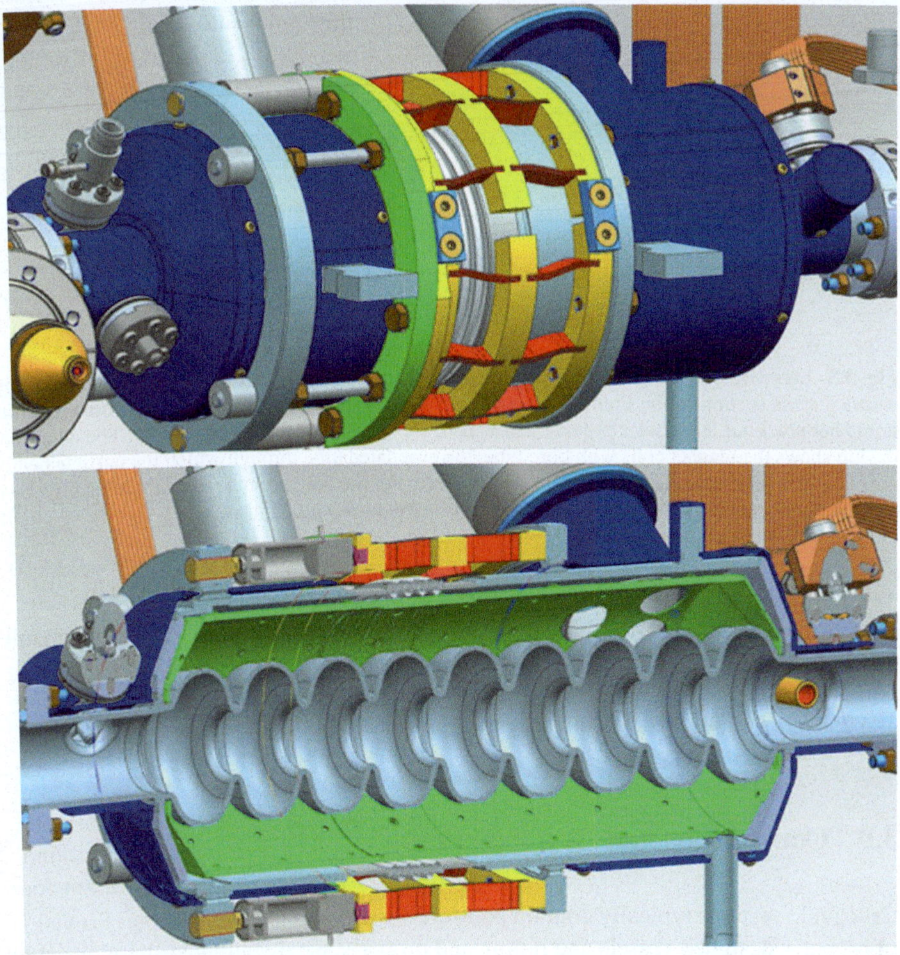

Fig. 3.7 Nine-cell RF cavity within helium vessel surrounded by perforated magnetic shield. Outer vessel CAD model and cross section from CAD model of assembly are shown. Area of holes in magnetic shield and helium space around magnetic shield within helium vessel were sized for venting flow with loss of cavity vacuum as well as for steady-state operation (*Images* Fermilab. *Magnetic shield design* Saravan K. Chandrasekaran, personal communication)

3.8 More General Cryogenic System Safety Reviews

A few review techniques which go beyond specifically pressure safety are worth briefly summarizing here. Those are Failure Mode and Effects Analysis (FMEA) and What-if Analysis. An FMEA considers the impact of failures of devices and instrumentation one at a time in the cryogenic system. This analysis may be generated by following a valve and instrument list and considering the consequences of failure for each device.

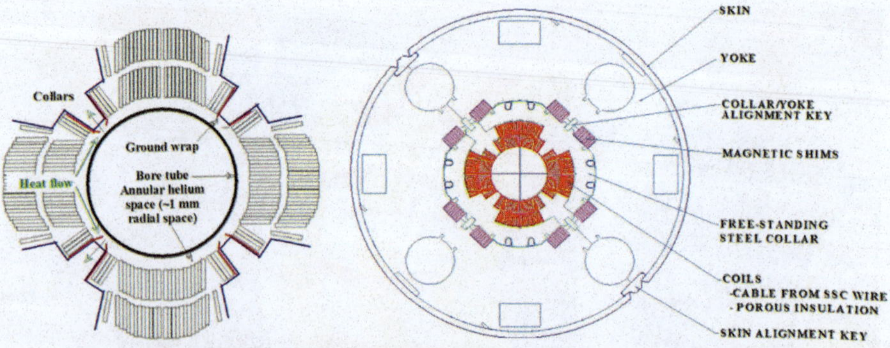

Fig. 3.8 Large Hadron Collider (LHC) final focus quadrupole cross section. The image on the right shows a cross section of the liquid helium-cooled portion of the magnet; the "skin" contains the liquid helium. Four large holes provide a flow path for the bulk of the liquid helium. The detail on the left shows the magnet coils with bore tube, the evacuated tube through which the particle beam passes. An annular space for helium of 1 mm radius under the coil, around the beam tube, provides a channel for heat transport and helium venting flow over the length of the magnet, about 6 m length in this case (*Image credits* Fermilab)

A What-if Analysis considers the consequences of larger events which may result in multiple failures such as a power outage sending some number of valves to their default positions. See Chap. 8, Approaches to Cryogenic Safety in Particle Accelerator Labs, for more information about FMEA and What-if methods.

3.9 Conclusions and Best Practices

Cryogenic systems typically contain pressure vessels and pressure piping. Provision of properly designed vessels and piping and properly sized venting systems are vital for system safety. Engineering for pressure safety in cryogenics, as for any pressure system, includes the following considerations:

- Select Maximum Allowable Working Pressures (MAWP) for vessels and piping based on system requirements and safety considerations.
- Design the vessels (pressure vessels and vacuum vessels) and piping for this MAWP using code standards or appropriate levels of safety as required in accordance with governing rules and laws. For U.S. Department of Energy Laboratories, this means a level of safety equivalent to the standard pressure codes.
- Evaluate all possible sources of pressure in the design, including maximum flow rates, temperatures, and pressure.
- Design venting systems (flow path to the relief device, relief device size, and any downstream ducting) to protect the vessels and piping against exceeding the MAWP.

- Document the analyses and designs showing that the above requirements have been met with clear, reviewable reports. Generation of these documents will begin as part of the design phase of the project and then remain as records helping to provide assurance of system safety.

References

1. Texas State Fire Marshal's Alert, February 22, 2006, University Campus Liquid Nitrogen Cylinder Explosion, http://www.tdi.texas.gov/fire/documents/fmred022206.pdf. Accessed 2018
2. ASME BPVC-VIII-1 2015—2015 BPVC Section VIII—Rules for Construction of Pressure Vessels—Division 1, and ASME BPVC-VIII-2 2015—2015 BPVC Section VIII—Rules for Construction of Pressure Vessels—Division 2—Alternative Rules
3. ASME BPVC-IID-C 2015/VOL 63 INTS—2015 BPVC Section II—Materials—Part D—Properties (Customary)
4. ASME Pressure Piping Code, B31.3, Process Piping
5. Pressure equipment standards and regulations for countries and regions outside of the United States may differ considerably in details for implementation. Examples include the Japanese Pressure Equipment Regulations, European Pressure Directive (PED)
6. Code of Federal Regulations, 10 CFR Part 851, "Worker Safety and Health Program"
7. SLAC ES&H Manual Chapter 14, "Pressure Safety"
8. Fermilab's ES&H Manual (FESHM) Chapter 5031
9. T. Boeckmann et al., *Experimental Tests of Fault Conditions During the Cryogenic Operation of a XFEL Prototype Cryomodule*. DESY 2008
10. M. Wiseman et al., Loss of cavity vacuum experiment at CEBAF. Adv. Cryog. Eng. **39**, 997 (1994)
11. E.G. Brentari et al., National Bureau of Standards Technical Note 317, in *Boiling Heat Transfer for Oxygen, Nitrogen, Hydrogen, and Helium.* 1965
12. American Petroleum Institute, *Recommended Practice for the Sizing, Selection, and Installation of Pressure-Relieving Systems in Refineries*, API RP 520, Part 1, 8th edn. (API, Washington, DC, 2008)
13. American Petroleum Institute, *521 Pressure-Relieving and Depressuring Systems*, 6th edn., American Petroleum Institute 2014
14. Compressed Gas Association, Inc. S-1.3, *Pressure Relief Device Standards Part 3—Stationary Storage Containers for Compressed Gases*, 8th edn. (CGA, Chantilly, VA) 2002
15. R. Byron Bird, W.E. Stewart, E.N. Lightfoot, *Transport Phenomena* (Wiley, New York, 1960)
16. Crane Technical Paper #410, *Flow of Fluids through Valves, Fittings, and Pipes* (Crane Co.) 1982 (updated editions available)
17. NBS Technical Note 631, *Thermophysical Properties of Helium-4 from 2 to 1500 K with Pressures to 1000 Atmospheres* (1972)
18. HEPAK, by Cryodata, Inc.
19. *Technology of Liquid Helium*, NBS Monograph 111
20. Anderson Greenwood, Crosby and Varec products, Technical publication No. TP-V300, https://www.emerson.com/documents/automation/pressure-relief-valve-engineering-handbook-en-us-3923290.pdf. Accessed 2018
21. D.A. Crowl, J.F. Louvar, Relief sizing, Chapter 10 in *Chemical Process Safety: Fundamentals with Applications*, 3rd edn. (Prentice Hall, Englewood Cliffs, NJ) 2012
22. R. Byrns, Y. Huang, J. Kerby, P. Lebrun, L. Morrison, T. Nicol, T. Peterson, R. Trant, R. van Weelderen, J. Zbasnik, The cryogenics of the LHC interaction region final focus superconducting magnets, in *Proceedings of the Seventeenth International Cryogenic Engineering Conference, ICEC 17*, Bournemouth, UK, 1998

23. P. Lebrun, T. Wahlstrom, R. van Weelderen, L. Williams, Investigation of quench pressure transients in the LHC superconducting magnets, in *Proceedings of the Fifteenth International Cryogenic Engineering Conference, ICEC 15*, Genova, Italy, 1994
24. J.E. Fesmire, S.D. Augustynowicz, C. Darve, Performance characterization of perforated multilayer insulation blankets, in *Proceedings of the 19th International Cryogenic Engineering Conference (ICEC19)*, Grenoble, France (2002), pp. 843–846
25. W. Lehman, G. Zahn, *Safety Aspects for LHe Cryostats and LHe Transport Containers* (ICEC7, London, 1978)
26. G. Cavallari et al., Pressure protection against vacuum failures on the cryostats for LEP SC cavities, in *4th Workshop on RF Superconductivity*, Tsukuba, Japan, 14–18 August 1989

Chapter 4
Oxygen Deficiency Hazards

Abstract Oxygen Deficiency Hazards (ODH) are ubiquitous in cryogenic systems. Such hazards can be both insidious and potentially fatal. This chapter describes the nature of the hazard including the physiological effects of oxygen depletion and the mitigations taken to reduce the hazard to acceptable levels. A detailed discussion of how to conduct an ODH Risk Analysis and apply mitigations is presented. The proper response to ODH incidents is discussed along with examples of recent studies, both numerical and experimental, of helium venting scenarios. A list of best practices is included. The chapter appendix contains equipment and human factor failure rates to assist in risk analysis along with a complete example ODH risk analysis.

4.1 Example Accident

In March of 1981, three technicians working at the Kennedy Space Center entered a compartment in the aft section of the space shuttle Columbia that had been purged with gaseous nitrogen. Due to a combination of poor communication and inadequate procedures, the technicians were unaware of the presence of an oxygen deficient environment in the compartment. All three technicians collapsed immediately. Two other workers entering the compartment in an attempt to rescue the first three also collapsed. Two of the three initial technicians died and one of the collapsed rescuers died several years later due to complications from the accident [1]. This example illustrates several common features of Oxygen Deficiency Hazards (ODH) accidents: a lack of awareness that the hazard exists, immediate incapacitation upon entering the ODH area, injury to would be rescuers and multiple fatalities.

4.2 Nature of the Hazard

One of the more insidious and dangerous hazards posed by cryogenic systems is the displacement of oxygen in the air by gases released from cryogens such as liquid nitrogen or liquid helium. This is known as an oxygen deficiency hazard (ODH).

T. J. Peterson and J. G. Weisend II, *Cryogenic Safety*, International Cryogenics Monograph Series, https://doi.org/10.1007/978-3-030-16508-6_4

Table 4.1 Effects of oxygen deficiency [8]

Volume% oxygen (at sea level)	Effect
17	Night vision reduced Increased breathing volume Accelerated heartbeat
16	Dizziness Reaction time for novel tasks doubled
15	Impaired attention Impaired judgment Impaired coordination Intermittent breathing Rapid fatigue Loss of muscle control
12	Very faulty judgment Very poor muscular coordination Loss of consciousness Permanent brain damage
10	Inability to move Nausea Vomiting
6	Spasmodic breathing Convulsive movements Death in 5–8 min

Oxygen constitutes 21% of normal air by volume. Oxygen is, of course, necessary for human life and an oxygen deficiency hazard is generally defined as any situation in which oxygen is 19.5% or less by volume at a barometric pressure of 1 bar. Table 4.1 shows the physiological effects of reduced oxygen as a function of percent. Without adequate oxygen, one can lose consciousness in a few seconds and die of asphyxiation in a few minutes. *It's critical to understand that at sufficiently low levels of oxygen, the first symptom is unconsciousness. It is this lack of warning that makes oxygen deficiency such a dangerous hazard.* Figure 4.1 shows the time of useful consciousness as a function of oxygen concentration. Note that at sufficiently low concentrations, unconsciousness occurs in less than 30 s with no warning.

One aspect of cryogenic fluids that helps create ODH situations is the very large volume ratio between a cryogenic liquid at its boiling point and the resultant gas at room temperature and pressure (see also Chap. 1). This is illustrated for common cryogens in Table 4.2. This volume ratio can easily exceed 700. Thus, a relatively small amount of cryogen can displace all the oxygen in an enclosed space. In fact, one of the most frequently overlooked hazards is the use of a standard 160 L LN2 dewar in a small lab. Oxygen deficiency hazards are not restricted to the use of cryogenic liquids. Sufficient amounts of inert gases such as nitrogen and helium stored at room temperature can pose a hazard if not properly handled and vented. Large quantities of inert gases are frequently stored under pressure in cryogenic facilities.

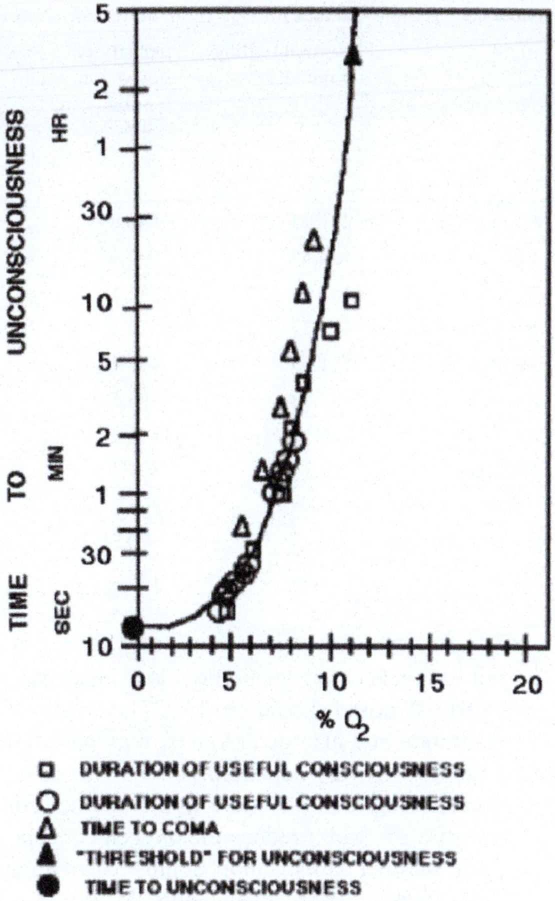

Fig. 4.1 Approximate time of useful consciousness for a seated subject at sea level versus % O_2. *Courtesy* B. Soyars, Fermilab [9]

Table 4.2 also shows the densities of these gases at 300 K and 1 bar. These may be compared to the average density of dry air (roughly 1.2 kg/m^3) under the same conditions. Thus, helium will tend to move upwards and concentrate near the ceilings while argon will tend to move down and concentrate in pits, trenches and basements. However, keep in mind that cold gas from a significant release will be much denser and in this case even helium may move to lower spaces. Also, in a significant enough leak, even warm helium may quickly render an entire space oxygen deficient.

4.3 Basics of ODH Safety

The steps for reducing oxygen deficiency hazards are:

1. Be aware of the nature of oxygen deficiency hazards. *Never use cryogenic liquids, regardless of the amount, without considering whether or not an oxygen deficient environment may result.*

Table 4.2 Volume changes for cryogenic fluids from normal boiling point to 300 K & 1 bar

Fluid	Normal boiling point (K)	Density of liquid at normal boiling point (kg/m^3)	Density of gas at 1 bar and 300 K (kg/m^3)	Volume of gas at 1 bar, 300 K/volume of liquid at normal boiling point
Propane	231.07	580.89	1.80	323
Ethane	184.55	543.97	1.22	446
Xenon	165.04	2942.1	5.29	556
Krypton	119.77	2416.3	3.40	711
Methane	111.63	422.42	0.64	660
Argon	87.28	1395.5	1.62	861
Oxygen	90.19	1142.2	1.3	879
Nitrogen	77.2	807.3	1.12	720
Neon	27.09	1205.2	0.81	1488
Hydrogen (Para)	20.23	70.85	0.081	875
Helium	4.222	125.2	0.16	783

2. For each use of cryogenic liquids or inert gases a formal written analysis of the ODH risk posed should be done. The details of this may vary from institution to institution and may be driven by regulatory requirements. The level of analysis required will vary depending on the situation. In some cases, a simple calculation showing that all liquid in a container converted to room temperature and pressure gas won't reduce the oxygen concentration below 19.5% may be sufficient. In other cases, a more sophisticated analysis may be needed. Examples of these approaches are given below. All analysis, even the simplest ones should be independently reviewed and documented.
3. Apply mitigations to reduce the risk of an ODH incident.
4. Have a plan for responding to emergencies and alarms.
5. Train all affected staff, including visitors on the nature of the hazard, safety requirements and how to respond to alarms.

4.4 Mitigations

Described below are a number of typical mitigations used to reduce the risk of ODH accidents.

The best solution is to mitigate or eliminate the hazard by design choices. These choices include:

1. Reduce inventory of cryogenic fluids & compressed gases.

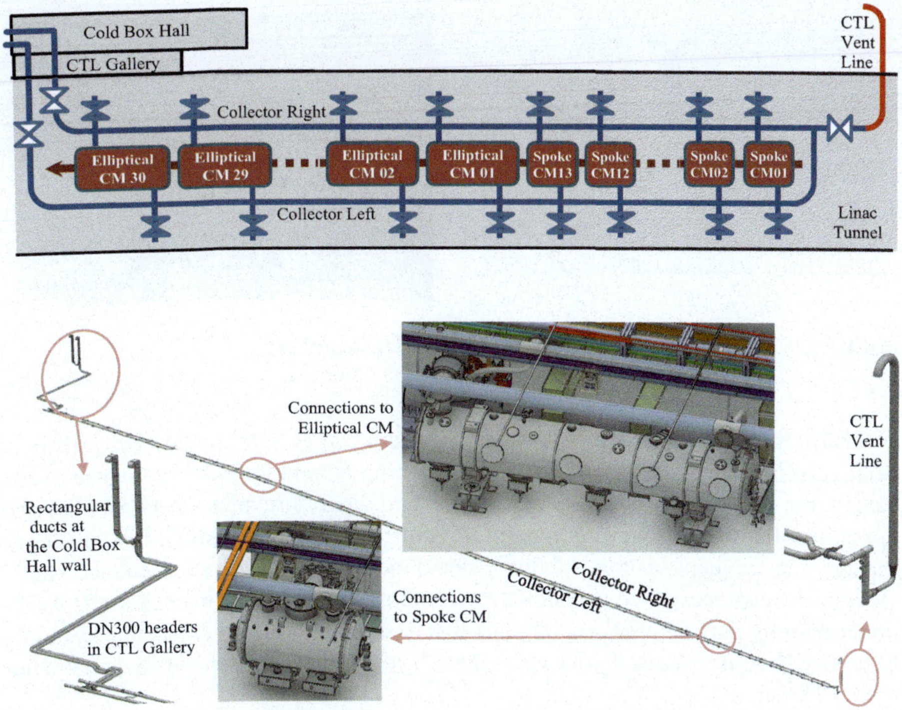

Fig. 4.2 Design of the safety helium vent collection header in the ESS tunnel [15]

- Use minimum amounts of cryogens or oxygen displacing gas
- Restricted Flow Orifaces (RFOs) passive devices used in conjunction with compressed gas systems to reduce the amount of oxygen displacing gas that can enter an area

2. Do not conduct cryogenic activities in small spaces
3. Do not transport cryogens in closed vehicles or in elevators with people
4. Do not use LN2 underground
5. Whenever feasible, vent relief valves into collection headers or to the outside of building (away from any air intakes or constricted areas) and not into interior spaces. An example of such a header employed in the European Spallation Source (ESS) tunnel is shown in Fig. 4.2
6. Where appropriate, employ lintels or other devices to direct released inert gases away from exits, escape routes or transportation aisles. Figure 4.3 shows an example of this approach employed at Jefferson Lab.

However, since with cryogenic systems it is generally impossible to guarantee that inert gases won't be released and cause an oxygen deficient hazard some of the mitigations below are used in addition.

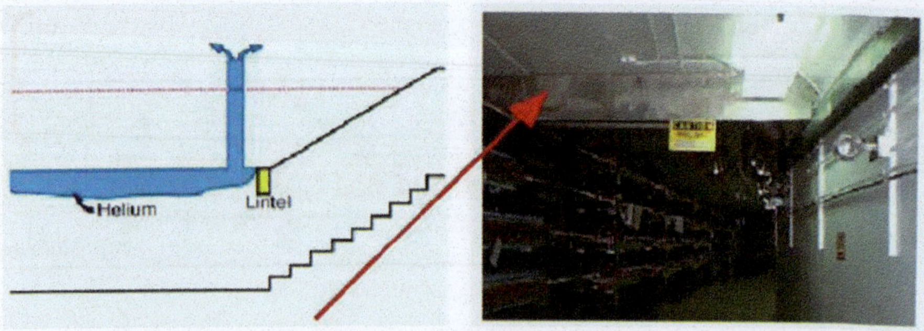

Fig. 4.3 Lintel installation in the Jefferson lab accelerator tunnel [8]

Training: Everyone who works in areas where Oxygen Deficiency Hazards may exist must be trained. At a minimum, the training should include the nature of the hazard, the proper response to ODH alarms (and what those alarms sound like) and operation of any emergency equipment if required. The training should include contractors and frequent visitors (delivery personnel, security and maintenance staff). Occasional visitors can be trained or continuously escorted by trained staff. Additional training may be required for staff that design and operate cryogenic systems. Everyone should be trained not to introduce cryogens into an area without analyzing the ODH risk.

Signs: These notify people of the hazard and proper response and indicate that only trained people are authorized to be there.

Fixed Oxygen Monitors and Alarms: Due to the insidious nature of ODH, active monitoring of the oxygen concentration in areas where ODH may occur is a common and very valuable mitigation. There are a number of reliable oxygen monitoring systems commercially available [2]. Some monitors have been shown [3] to not indicate a low oxygen environment if helium is the inert gas displacing the air and care must be taken in selecting the monitoring system. Additional guidelines include:

1. Monitors are generally set to alarm at the regulatory limits for oxygen deficiency.
2. Alarms should be both audio and visual (this is particularly important in high noise areas such as compressor facilities) and should be directly connected to emergency responders or at minimum to a central control room.
3. Given that many people ignore fire alarms without some secondary indication of fire or smoke, it is highly advisable to have a separate, distinct ODH alarm.
4. ODH alarms should also be installed at the entrances to potential ODH areas to prevent people from entering spaces with low oxygen levels.
5. Oxygen monitors and alarms should be powered by uninterruptable power supplies and procedures should be established for safely entering a space whose oxygen monitoring system has shut down for any reason.
6. The design and implementation of oxygen monitoring systems should be independently reviewed.

A description of an ODH system designed for the ESS project is given in reference [4].

Personal Oxygen Monitors: These monitors, worn on one's body, provide additional protection, particularly in large facilities where there may be localized areas of low oxygen that might be missed by fixed monitors. Suppliers of these may also be found in reference [2].

Two Person Rule: Two people working together in constant communication. This doubles the number of people to observe alarms or unsafe conditions as well as allowing someone to seek help in the event of an accident with the other person.

Third Person Observer: This is a third person (assuming the 2 person rule is in place) that observes the work from outside the ODH area. This person can summon help in the case of an accident.

Ventilation and Other Active Systems: Oxygen deficiency hazards can be reduced by ensuring an adequate amount of ventilation or by having systems that close off cryogen supplies in the event of a low oxygen indication or other signs of a leak. While this solution is quite attractive, if you depend on ventilation or other active systems for safety, then these are now safety systems. Issues of system reliability, downtime for maintenance, uninterruptable power supplies must all be addressed in order for these systems to mitigate oxygen deficiency hazards.

Emergency Escape Packs: In facilities such as tunnels, where the distance between exits is may be large, staff working in potential ODH areas may be required to carry with them escape packs that provide supplementary air sufficient for them to safely leave an ODH environment. Proper training is vital for these systems to be effective.

Work Restrictions: In some cases the hazards, are so significant that staff are prohibited from entering potential Oxygen Deficiency areas under certain conditions, such as magnets being under power or during cryogenic cool down or warm up cycles. Such work restrictions have been established as CERN [5] and similar restrictions are envisioned for work in the ESS tunnel. Work restrictions may also limit the type of work to be conducted when large amounts of cryogenic fluids are present. For example, ESS will likely not permit heavy transport activities in the tunnel, which could potentially damage the cryomodules while the cryomodules contain liquid helium.

4.5 ODH Risk Analysis

In many institutions, including the European Spallation Source [6], Fermilab [7] and the SLAC National Accelerator Laboratory [8] the formal Oxygen Deficiency Hazard risk analysis is a two-step process. A simple scaling analysis followed, if needed, by a more formal Risk Assessment.

4.5.1 Scaling Analysis

First a simple calculation is used to determine the extent of the hazard. In the case of SLAC, this done by using the following process:

In the case of the source of cryogens or inert gas being inside the room (for example in storage dewar or gas cylinder) calculate:

$$Oxygen\ Level(\%) = \frac{21(V_R - V_C)}{V_R} \tag{4.1}$$

where V_R is the volume of the room and V_C is the volume of the inert gas at room temperature and pressure.

In the case where the source is outside the room and flowing into the system a given rate (for example flow from a helium compressor) calculate:

$$Oxygen\ Level(\%) = \frac{21(V_R - Q)}{V_R} \tag{4.2}$$

where V_R is the volume of the room and Q is the volumetric flow rate of the inert gas at room temperature and pressure. Note that this calculation assumes one exchange of room air per hour.

In the SLAC ODH policy, if the oxygen levels calculated in Eqs. 4.1 or 4.2 are less than 19.5% under normal operating conditions or less than 18% under abnormal conditions then a more sophisticated risk assessment (see below) is required. Similar rules apply at ESS and CERN. In all these institutions, the simple scaling analysis must be documented and approved by an independent reviewer. An independent review and approval of any ODH analysis is a critical for safety. An example of the form used at SLAC for documenting this scaling analysis is given in the chapter appendix.

Keep in mind the underlying assumption of uniform mixing. Be aware of helium being trapped at a high level, argon gas concentrated in pits or trenches and the possible effect of gas colder than 300 K being released. Any doubt about the applicability of Eqs. 4.1 and 4.2 to the situation under review should result in applying the risk assessment technique described below.

Example Calculation

Assume you have a small lab that is 2.5 m by 2.5 m by 3 m in size. Would a standard 160 L dewar of liquid nitrogen pose an ODH risk?

Answer: The volume of the room equals 2.5 m by 2.5 m by 3 m or 18.75 m^3 which equals 18,750 L. The maximum volume of LN2 that can be spilled is 160 L but that is equivalent (see Table 4.2) to 720 × 160 = 115,200 L of nitrogen at 300 K and 1 bar. Since this amount is much bigger than the room volume we can see that all the oxygen will be displaced and a significant ODH risk is present.

Applying some sort of engineering solution such as moving to a larger room or at least moving the dewar out of the room and using a flow limiter on the connecting hose is needed. The analysis should be repeated after the design is changed and a more rigorous risk assessment (see below) done if required by Eqs. 4.1 or 4.2.

4.5.2 *Risk Assessment*

A more detailed and formal way to analyze the oxygen deficiency hazard is to calculate a probabilistic fatality rate per hour and then associate that risk with different ODH classes. Each ODH class is linked with a set of required mitigations that are meant to reduce the risk to acceptable levels. The total fatality rate per hour is found by summing all the possible ODH risk hazards

$$\Phi = \sum_{i=1}^{n} P_i F_i \tag{4.3}$$

where

Φ the ODH fatality rate (per hour)
P_i the expected rate of the i-th event (per hour), and
F_i the probability of a fatality due to event i.

The value of P_i for each event is generally taken from institutional or industry experience. The chapter appendix contains reference values for equipment failure rates and human error rates used at both ESS and Fermilab.

The probability of a fatality resulting from a given event is also a judgment based on experience. One approach is to link F_i with the oxygen concentration that results form the ODH event. An example of this (from ESS) is shown in Fig. 4.4. Note that at above 18% oxygen the fatality probability is taken to be zero and that below 8.8% the fatality probability is taken to 1. This latter choice results from the rapid onset of unconsciousness at these oxygen concentrations.

The oxygen concentration that results from a given event depends on the details of the event itself and the surrounding environment. This includes such factors as the rate of release of the cryogenic gas or liquid, the size of the space involved, the presence of any ventilation and the degree of mixing of the released gas with the air. When calculating the resulting oxygen concentration care must be taken to use conservative assumptions and fully document them. Fermilab [9] has developed a number of useful

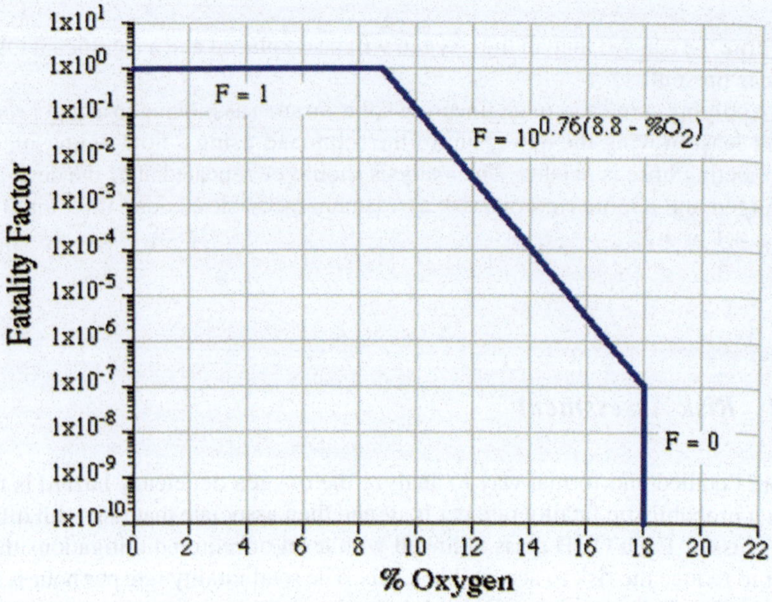

Fig. 4.4 Fatality factor F_i versus the lowest attainable oxygen concentration that can result from a given event [6]

Table 4.3 Typical equations for estimating Oxygen concentration resulting from the release of an inert gas [9]

#	Time	Mixing	Ventilation	Type	Diff. eqs.	Concentration
A	During release	Perfect mixing	Into confined volume		$V\frac{dC}{dt} = 0.21Q - (R+Q)C$	$C(t) = \left(\frac{0.21}{Q+R}\right)\left[Q + R\,e^{-\left(\frac{Q+R}{V}\right)t}\right]$
B	During release	Perfect mixing	From confined volume	Ventilation rate greater than spill	$V\frac{dC}{dt} = 0.21(Q - R) - QC$	$C(t) = 0.21\left(1 - \frac{R}{Q}\left(1 - e^{-\frac{Q}{V}t}\right)\right)$
C	During release	Perfect mixing	From confined volume	Ventilation rate less than spill	$V\frac{dC}{dt} = -RC$	$C(t) = 0.21e^{-\frac{R}{V}t}$
D	After release	Perfect mixing			$V\frac{dC}{dt} = 0.21Q - QC$	$C(t) = 0.21 - [0.21 - C_r(t_e)]e^{-\frac{Q}{V}(t-t_e)}$

equations for estimating oxygen concentrations based on certain conditions. Table 4.3 summarizes these equations. See reference [9] for additional details. Note that all these equations assume perfect mixing. Depending on the specific scenario under investigation, other equations may be more appropriate.

Definitions:

C oxygen concentration [%].
C_r oxygen concentration during the release [%].
C_e oxygen concentration after the release has ended [%].
Q ventilation rate of fan(s) [m^3/s].
R spill rate into confined volume [m^3/s].
t time [s] (beginning of release is at $t = 0$).
t_e time when release has ended [s].
V confined volume [m^3].

In summary, for a given system, the rate of each possible event is determined and multiplied by the corresponding fatality probability. The fatality rates for all the events are then summed together to produce a total ODH fatality rate. Once the total ODH fatality rate per hour is calculated, the ODH class can then be determined. The definition of such classes varies from institution to institution and Table 4.4, from ESS, is a typical example. Many institutions would not allow situations higher than Class 2 to exist but would instead require redesign of the system (for example by reducing cryogen inventory or relocating it outdoors) to reduce the ODH class to level 2 or lower. The remaining classes are then associated with required mitigations that have the effect of further reducing the risk. Table 4.5, from ESS, is a typical example of linking mitigations with ODH classes. Note in Table 4.5 that ESS policy is not to permit ODH classes higher than 2.

The risk assessment approach, used by a number of laboratories in the world has the advantage of attempting to capture all the possible ODH events and determining the size of the events hazard. While quantitative, this approach still requires significant judgment and estimation. Again, an independent review of the ODH Risk Assessment is necessary to maximize safety.

In the case of ESS, our preliminary ODH analysis showed that the surface buildings containing cryogenic systems could be rated as a Class 0. However, given the large amounts of inert gases present, the ESS ODH Safety Committee decided to require a fixed ODH monitoring system for these buildings as an additional mitigation.

Example Calculation. The best way to illustrate the use of these risk assessments is by a real world example. Such an example is given in the Chapter Appendix. This detailed example concerns the ODH analysis of a small lab using LN2 and LHe. It's

Table 4.4 An example of ODH classes [6]

Class	$\Phi(Hr^{-1})$
0	$<10^{-7}$
1	$>10^{-7}$ but $<10^{-5}$
2	$>10^{-5}$ but $<10^{-3}$
Forbidden	$>10^{-3}$

Table 4.5 ODH classes and required mitigations from the ESS ODH policy [6]

ODH class	0	1	2
Technical safety measures			
Warning signs	X	X	X
Ventilation		*	*
Area (fixed) oxygen monitoring	*	X	X
Organizational safety measures			
Medical approval as ODH qualified		*	*
ODH training (e-learning)	X	X	X
Personal oxygen monitor		X	X
Self-rescue mask		*	*
Presence of minimum 2 persons			X
Administrative safety measures			
Access restricted to authorized personnel only		X	X
Emergency procedure		X	X
Operating procedure	X	X	X

X means the mitigation is required
*means that the mitigation to be evaluated case by case with the help of cryogenic safety committee or the ESS environmental health and safety division

indicative of the level of detail and thought required for the ODH risk analysis of such a system.

Similar approaches to determining and mitigating ODH risk are carried out at CERN [7], Argonne National Lab [10] and Jefferson Lab [11]. The key point is to rigorously determine the hazard, considering all possible events, and then apply mitigations to reduce the risk to acceptable levels. A formal review of the analysis and mitigations is vital for this approach to succeed.

4.6 Proper Response to ODH Alarms and Events

In the event of an ODH alarm, people should immediately leave the potential ODH area, avoiding walking through obvious vapor clouds. Re-entry into the area to inspect or fix the problem should only be done by specially trained staff with supplemental oxygen if needed. Likewise, rescue of people collapsed during an ODH event must only be down by properly trained emergency personnel, such as the fire brigade. Rescue attempts by untrained staff can easily lead to multiple causalities.

4.7 Helium Vapor Release Studies and Numerical Modeling

In large and complex systems, it may be useful to conduct experiments to further understand the ODH risks and the impact of associated mitigations. By far, the most valuable approach is to conduct full-scale tests in which relevant amounts of inert gas or liquid are released into the physical space under study. Such tests are time consuming, expensive and have to be carefully designed to yield useful information. Thus, they are generally restricted to situations in which a specific question is being asked or in systems with potentially very large Oxygen Deficiency Hazards and limited escape routes.

Despite their complexity, release studies have been conducted over the years and have generated important results. An experiment carried out at Jefferson Lab to verify the design of a lintel system to protect exit stairs from ODH situations had the added benefit of demonstrating, much to everyone's surprise, that some of the ODH monitors in use at the time did not respond appropriately when the inert gas was helium [2].

A more recent example is a series of full scale venting tests carried out in the CERN Large Hadron Collider (LHC) tunnel. These tests [5] released up to 1000 L of liquid helium into the LHC tunnel at rates of 0.1, 0.3 and 1 kg/s. The experiment measured temperature, flow rate and oxygen concentration at different locations along the tunnel at either side of the helium release. Video footage of the releases and propagation of the vapor cloud were also made. Figure 4.5 is an example of some of the data taken during this experiment. Another significant result was that even in areas outside the visible vapor cloud the oxygen concentration dropped below 19.5% due to the presence of warm helium gas from the spill. This illustrates the necessity of having a reliable ODH monitoring system in place and not relying on visual cues to determine the hazard.

These measurements resulted in new work rules at CERN prohibiting tunnel access during cool down and warm up of the system and overall severally limiting access to system experts whenever liquid helium is present in the accelerator. Overall, CERN has limited tunnel access to ensure that staff is not exposed to helium releases greater than 0.1 kg/s.

Numerical modeling of potential ODH events can be used as a supplement or alternative to experimental measurements. However, releases of inert gases due to various accident scenarios tend to be very dynamic events with rapidly changing temperatures and pressures and thus rapidly changing fluid properties. In addition, a number of assumptions typically have to be made regarding the amount of gas released, the temperature of the gas as it enters the room and so on. These two features of dynamic behavior and underlying assumptions mean that a solely numerical analysis may not be very accurate. A numerical model that is not verified by measured data should not be used to justify a lower ODH risk.

Numerical models can be used for scaling purposes and certainly are valuable in showing how bad a potential ODH situation may be. Figure 4.6 shows some results of a computational fluid dynamics model of helium venting in the ESS tunnel. While we

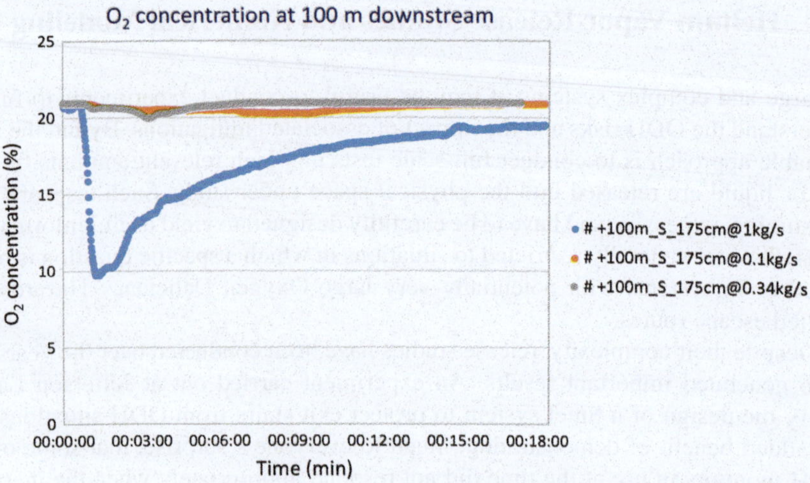

Fig. 4.5 Measured oxygen concentration versus time and helium release rate at a given location in the LHC tunnel relative to the helium spill (Courtesy T. Koettig & D. Delikaris–CERN)

did not necessarily believe the absolute values of the resulting oxygen concentrations and temperatures, this work certainly illustrated the need for the helium collector shown in Fig. 4.2. It also motivated reinforcing the hinges of the tunnel exit doors so that the overpressure from venting into the tunnel will not render the doors inoperable.

Numerical models that are benchmarked by experimental measurements (preferably full-scale tests) have significantly more value. The CERN tunnel tests described above were partly motivated by a desire to test an existing numerical model. Additional example of such a combination of modeling and experiments are given in references [12–14].

4.8 Best Practices

1. Always evaluate the risk of an oxygen deficiency hazard whenever dealing with cryogenic fluids or compressed gases, no matter how small the amount.
2. Develop a well understood and documented approach to evaluating and mitigating oxygen deficiency hazards.
3. Consider at the beginning of the system design, choices that minimize or eliminate oxygen deficiency hazards. These may include: reducing the cryogenic inventory, limiting the amount of inventory that can be released in an accident, venting relief systems outdoors, maximizing interior space in which cryogenic systems are contained and designing tunnels with sufficient exits and ventilation to reduce ODH risk.
4. Have all ODH analyses, even simple ones, independently reviewed.

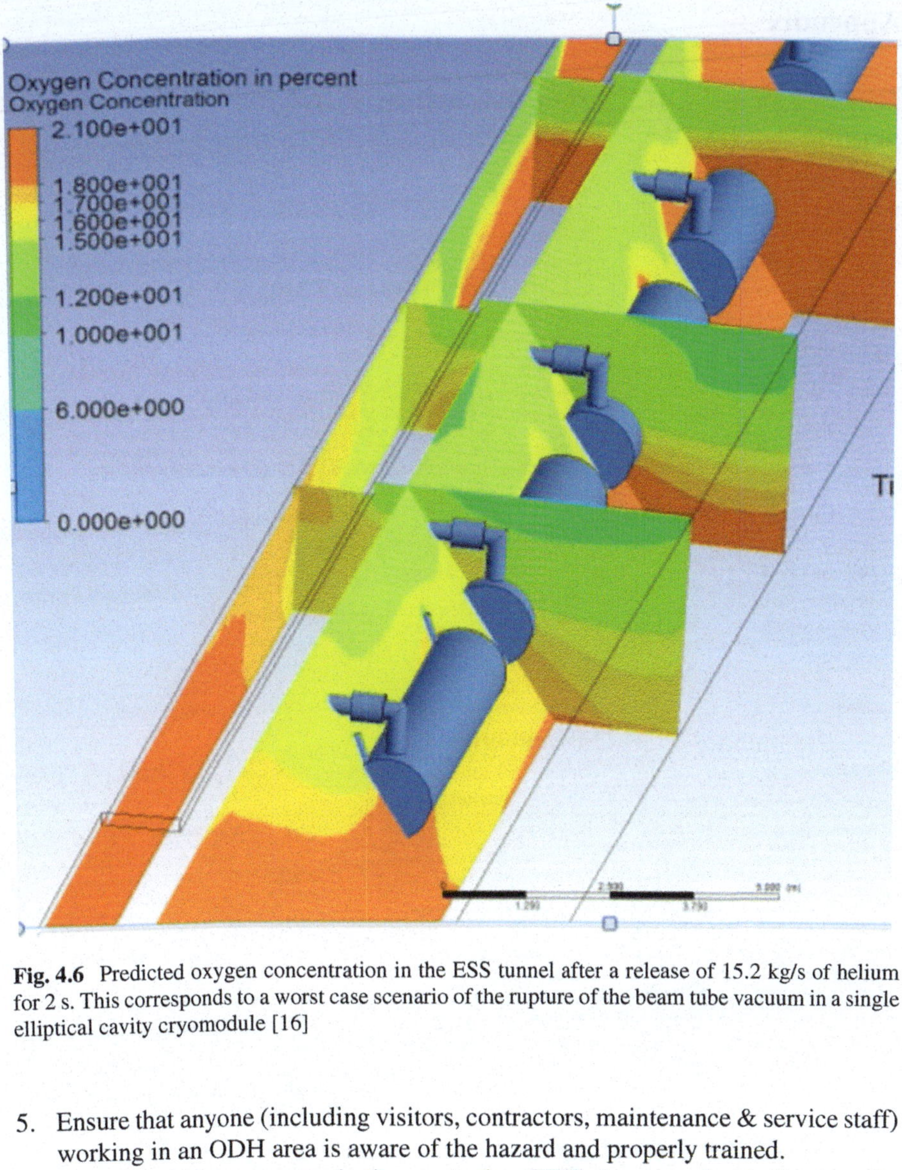

Fig. 4.6 Predicted oxygen concentration in the ESS tunnel after a release of 15.2 kg/s of helium for 2 s. This corresponds to a worst case scenario of the rupture of the beam tube vacuum in a single elliptical cavity cryomodule [16]

5. Ensure that anyone (including visitors, contractors, maintenance & service staff) working in an ODH area is aware of the hazard and properly trained.
6. Immediately leave an area in the event of an ODH alarm or other indication of a leak such as a vapor cloud. Do not walk through vapor clouds while exiting.
7. Never enter an area in which an ODH alarm has sounded or that is otherwise thought to be oxygen deficient unless you have been properly trained and equipped for such entries.
8. Do not transport even small amounts of cryogenic fluids inside cars or elevators.

Appendix

Form Used at SLAC National Accelerator Lab for ODH Scaling Analysis

SLAC
NATIONAL ACCELERATOR LABORATORY
ENVIRONMENT, SAFETY & HEALTH DIVISION

Chapter 36: Cryogenic and Oxygen Deficiency Hazard Safety

ODH Safety Review Form

Product ID: 669 | Revision ID: 1942 | Date Published: 21 June 2017 | Date Effective: 21 June 2017
URL: http://www-group.slac.stanford.edu/esh/eshmanual/references/cryogenicsFormODHReview.pdf | docx

This form is used to document the safety review required before introducing oxygen-displacing gases, including cryogens, into a work area or changing the existing use of such gases (for example, by adding or modifying systems, changing operations, or changing the quantity of gases used). The form is to be completed by the responsible person for the activity, reviewed by his or her ESH coordinator, and approved by the cryogenic and oxygen deficiency hazard (ODH) safety program manager. The completed form is to be maintained by the program manager and a copy kept by the responsible person (see Cryogenic and Oxygen Deficiency Hazard Safety: ODH Safety Review Procedure [SLAC-I-730-0A06C-001]).

1 General Information

Preparer		Location (bldg/rm/area)	
Responsible person		Directorate	
Description of system			
Gas to be introduced		Type of hazard	
Gas source		Additional comments *(see back for add. space)*	
Attachments	☐ Room floor plan showing location of ODH (required)	☐ Risk assessment (if required, see below)	

2 Preliminary ODH Calculation

If source is INSIDE proposed ODH area (e.g., dewar, tank)		If source is OUTSIDE proposed ODH area (e.g., from a compressor, pump)	
Total volume of room [a], V_R(ft³) =		Total volume of room, V_R(ft³) =	
Volume of gas at room temperature and pressure, V_0(ft³) =		Volume of flow rate of gas into room at room temperature and pressure, Q(ft³/h) =	
Calculate oxygen level, $21\frac{(V_R - V_0)}{V_R}$ =		Calculate oxygen level [b], $21\frac{(V_R - Q)}{V_R}$ =	
		Will ventilation be maintained during building power failure? [c]	☐ Yes ☐ No

a Room volume is calculated as length x width x ceiling height. Do not reduce room volume to account for room contents (cabinetry, machinery).

b This calculation assumes one air change per hour into the space.

c If there is no positive ventilation during a power failure, further evaluation is required to show that the area has sufficient passive ventilation or other mitigations are in place to assign an ODH 0 classification to the space.

If resulting oxygen level is ≥ 19.5% normal ops or ≥ 18% during system upset	Sign form and submit for approval
If resulting oxygen level is < 19.5% normal ops or < 18% during system upset	Conduct a risk assessment (see ODH Safety Review Procedure, Section 2.2) and complete Part 4, "Additional Information", then sign form and submit for approval

3 Approvals

Person	Name	Signature	Date
Responsible person			
ESH coordinator			
Cryogenic and ODH safety program manager			
ODH classification			
Comments			

21 June 2017 SLAC-I-730-0A06J-001-R003 1 of 2

4 Additional Information for Risk Assessment

To be completed by responsible person. List all that apply and be as specific as possible; indicate whether controls are existing or planned.	
Engineering Controls *Example:* *Mechanical ventilation* *Fume hoods* *Valves* *Critical orifices*	
Administrative Controls *Example:* *Training required* *Standard protective measures* *Work control documents*	
Attachments *List all that apply:* *Communications* *Risk assessment* *Hazard analysis*	

5 Additional Comments

General Information	
Approval	
Risk Assessment	

Equipment Failure Rates Estimates

System[a]	Failure mode	Failure rate
Compressor (two-stage Mycom)	Leak	5×10^{-6}/h
	Component rupture	3×10^{-7}/h
Dewar	Loss of vacuum	1×10^{-6}/h
Electrical power failure (unplanned)	Time rate	1×10^{-4}/h
	Demand rate	3×10^{-4}/D
	Time off	1 h
Fluid line (cryogenic)	Leak	5×10^{-7}/h
	Rupture	2×10^{-8}/h
Cryogenic magnet (powered, unmanned)	Rupture	2×10^{-7}/h
Cryogenic magnet (not powered, manned)	Rupture	2×10^{-8}/h
Header piping assembly	Rupture	1×10^{-8}/h
U-Tube change (cryogen release)	Small event	3×10^{-2}/D
	Large event	1×10^{-3}/D
Batteries, Power (UPC) supplies	No output	3×10^{-6}/h
Circuit Breakers	Failure to operate	1×10^{-3}/D
	Premature transfer	1×10^{-6}/h
DIESEL (complete plant)	Failure to start on demand	3×10^{-2}/D
Emergency run loads	Failure to run	3×10^{-3}/h
Engine only	Failure to run	3×10^{-4}/h
Electric motors	Failure to start on demand	3×10^{-4}/D
	Failure to run—normal	1×10^{-5}/h
	Failure to run—extreme	1×10^{-3}/h
Fans (fan, motor & starter)	Failure to run	9×10^{-6}/h
Fuses	Premature open	1×10^{-6}/h
	Failure to open	1×10^{-5}/D
Flanges	Leak, 10 mm^2 opening	4×10^{-7}/h
With reinforced & preformed gaskets	Rupture	1×10^{-9}/h
Flanges with packing or soft gaskets	Leak, 10 mm^2 opening	4×10^{-7}/h
	Packing blowout	3×10^{-8}/h

(continued)

(continued)

System[a]	Failure mode	Failure rate
	Rupture	1×10^{-9}/h
Instrumentation (amplification, transducers, calibration, combination)	Failure to operate	1×10^{-6}/h
	Shifts	3×10^{-5}/h
Motorized louver	Failure in continuous operation	3×10^{-7}/h
Piping	Small leak 10 mm^2	1×10^{-9}/h m
	Large leak 1000 mm^2	1×10^{-10}/h m
	Rupture	3×10^{-11}/h m
Piping weld	Small leak 10 mm^2	2×10^{-11}(D/t)/h m
D = diameter	Large leak 1000 mm^2	2×10^{-12} (D/t)/h m
t = wall thickness	Rupture	6×10^{-13} (D/t)/h m
Pumps	Failure to start on demand	1×10^{-3}/D
	Failure to run—normal	3×10^{-5}/h
	Failure to run—extreme	1×10^{-3}/h
Relays	Failure to energize	1×10^{-4}/D
	Failure no contact to close	3×10^{-7}/h
	Short	1×10^{-8}/h
	Open NC contact	1×10^{-7}/h
Solid state Hi Pwr application	Fails to function	3×10^{-6}/h
	Shorts	1×10^{-6}/h
Solid state Low Pwr application	Fails to function	1×10^{-6}/h
	Shorts	1×10^{-7}/h
Switches	Limit: fail to operate	3×10^{-4}/D
	Torque: fail to operate	1×10^{-4}/D
	Pressure: fail to operate	1×10^{-4}/D
	Manual: fail to transmit	1×10^{-5}/D
	Contact shorts	1×10^{-8}/h
Transformers	Open contact	1×10^{-6}/h
	Short contact	1×10^{-6}/h
Valves (motor operated)	Fails to operate (plug)	1×10^{-3}/D

(continued)

(continued)

System[a]	Failure mode	Failure rate
	Fails to remain open	1×10^{-4}/D
	External leak	1×10^{-8}/h
	Rupture	5×10^{-10}/h
Valves (solenoid operated)	Fails to operate	1×10^{-3}/D
Valves (air operated)	Fails to operate	3×10^{-4}/D
	Fails to remain open	1×10^{-4}/D
	External leak	1×10^{-8}/h
	Rupture	5×10^{-10}/h
Valves (check)	Fails to open	1×10^{-4}/D
	Reverse leak	3×10^{-7}/D
	External leak	1×10^{-8}/h
	Rupture	5×10^{-10}/h
Valves (orifices, flow meter)	Rupture	1×10^{-8}/D
Valves (manual)	Fails to remain open (plug)	1×10^{-4}/D
	External leak	1×10^{-8}/h
	Rupture	5×10^{-10}/h
Valves (relief)	Fails to open	1×10^{-5}/D
	Premature open	1×10^{-5}/h
Vessels (pressure)	Small leak, 10 mm^2	8×10^{-8}/h
	Disruptive failure	5×10^{-9}/h
Wires	Open	3×10^{-6}/h
	Short to GND	3×10^{-7}/h
	Short to Pwr	1×10^{-8}/h

[a]From *ESS Guideline for Oxygen Deficiency Hazard (ODH)*, ESS-0038692 (2016)

Human Error Rate Estimates

Estimated error rate D^{-1}	Activity[a]
1×10^{-3}	Selection of a switch (or pair of switches) dissimilar in shape or location to the desired switch (or pair of switches), assuming no decision error For example, operator actuates large handled switch rather than small switch
3×10^{-3}	General human error of commission, e.g., misreading label and therefore selecting wrong switch
1×10^{-2}	General human error of omission where there is no display in the control room of the status of the item omitted, e.g., failure to return manually operated test valve to proper configuration after maintenance
3×10^{-3}	Errors of omission, where the items being omitted are embedded in a procedure rather than at the end as above
1/x	Given that an operator is reaching for an incorrect switch (or pair of switches), he selects a particular similar appearing switch (or pair of switches), where x = the number of incorrect switches (or pair of switches) adjacent to the desired switch (or pair of switches). The 1/x applies up to 5 or 6 items. After that point the error rate would be lower because the operator would take more time to search. With up to 5 or 6 items he doesn't expect to be wrong and therefore is more likely to do less deliberate searching
1×10^{-1}	Monitor or inspector fails to recognize initial error by operator. Note: With continuing feedback of the error on the annunciator panel, the high error rate would not apply
1×10^{-1}	Personnel on different work shift fail to check condition of hardware unless required by check or written directive
5×10^{-1}	Monitor fails to detect undesired position of valves, etc., during general walk-around inspection, assuming no check list is used
0.2–0.3	General error rate given very high stress levels where dangerous activities are occurring rapidly
$2^{(n-1)}$ x	Given severe time stress, as in trying to compensate for an error made in an emergency situation, the initial error rate, x, for an activity doubles for each attempt, n, after a previous incorrect attempt, until the limiting condition of an error rate of 1.0 is reached or until time runs out. This limiting condition corresponds to an individual's becoming completely disorganized or ineffective

[a]From *ESS Guideline for Oxygen Deficiency Hazard (ODH)*, ESS-0038692 (2016)

Example: Helium and Nitrogen ODH Analysis for a Small Laboratory

Introduction

As an example of a simple Oxygen Deficiency Hazards analysis, we consider here ODH potential in a laboratory room which uses small volumes of liquid helium and liquid nitrogen for calibration of instrumentation and various other tests. The liquid helium is provided via transfers from 500 L commercial helium storage dewars, and the liquid nitrogen from 150 L commercial dewars. This analysis examines the oxygen deficiency hazard due to a release of cryogens in the laboratory.

Room Data

Lab floor area (= ceiling area) = 56 × 26 ft^2 – 11 × 11 ft^2 (subtract for a separate small room in one corner) = 1335 ft^2 = 124 m^2

Ceiling height = 13.75 ft = 4.2 m

Room volume = 18,400 ft^3 = 520 m^3.

Helium ODH Considerations

The maximum liquid helium container volume is 500 L. If released and warmed to room temperature this would fill 377 m^3, 72% of the room, down to a little over a meter off the floor if unmixed with the air. If mixed with the remaining air in the room, the concentration of oxygen would be 6%.

Therefore assume for helium:

(1) The release of the entire contents of a 500 L helium dewar results in a fatality (It is always full and all released).

(2) The rupture of the dewar insulating vacuum or any vessel or line attached to the dewar releases the entire contents of the dewar.

(3) A "large event" during a U-tube change is a release of the entire contents. From Table I in Fermilab's ODH standard [9], we have the following relevant probabilities:
   - dewar loss of vacuum $1 \times 10^{-6}/\text{h}$
   - cryogenic fluid line leak or rupture $5 \times 10^{-7}/\text{h}$
   - U-tube change release, large event $1 \times 10^{-3}/\text{demand}$

The lab could have as many as two 500 L helium dewars in it at once, each connected to a helium reservoir. Counting each reservoir as a dewar (connected via a transfer line so able to spill 500 L) the total probability of a dewar leak or rupture is 4×10^{-6}/h. There may be a transfer line to each reservoir and a vent line from each reservoir, for a total of 4 lines connected to a 500 L source, so the total probability of a cryogenic fluid line leak or rupture is 2×10^{-6}/h. Workers report 41 liquid helium transfers per year. The 41 LHe transfers per year entail 82 U-tube operations per year (counting stinging and removing as separate). So the total probability of a release (large event) during a U-tube change is 84×10^{-3}/year $= 9.6 \times 10^{-6}$/h.

Assuming every line leak or rupture, dewar loss of vacuum, or large U-tube release results in venting a full storage dewar, the total probability of the release of the entire contents of a 500 L helium dewar is $(4 + 2 + 9.6) \times 10^{-6}$/h $= 1.56 \times 10^{-5}$/h. Since a fatality factor of 1 has been assumed for these events, the ODH hazard class would be 2. But now consider ventilation.

Suppose a vent fan is connected to an ODH sensor, which is fail-safe, but the vent fan fails to start 5.6×10^{-3}/demand [9]. If this vent fan is sized to prevent the helium from accumulating in the engineering lab in the event of a 500 L release, then only when there is a release AND the fan fails to run is there a fatality. The probability of this occurring is $(5.6 \times 10^{-3}) \times (1.56 \times 10^{-5}/\text{h}) = 8.7 \times 10^{-8}$/h. This situation is now ODH 0 with very conservative assumptions regarding consequences of failures.

The above analysis narrative is summarized in Table 4.6.

Helium Release Rate and Required Fan Size

The maximum flow rate from a 500 L helium storage dewar would result from a loss of insulating vacuum to air and the resultant air condensation on the inner vessel. The venting could occur from a relief valve or another vent valve on the vessel if it

Table 4.6 Small laboratory helium ODH analysis summary

Device or action	Failure probability	Number of devices or actions	Total release probability	Probability vent fan fails to start	Release probability AND vent fan failure
Dewar loss of vacuum	1×10^{-6}/h	4	4×10^{-6}/h	5.6×10^{-3} per demand	2.24×10^{-8}/h
Line leak or rupture	5×10^{-7}/h	4	2×10^{-6}/h	5.6×10^{-3} per demand	1.12×10^{-8}/h
U-tube exchange helium release	1×10^{-3} per demand	82/year	9.6×10^{-6}/h	5.6×10^{-3} per demand	5.38×10^{-8}/h
			No ventilation		With ventilation
Total fatality rate assuming each fatality factor = 1			1.56×10^{-5}/h		8.74×10^{-8}/h
ODH rating			ODH 2		ODH 0

happened to be open. We will assume that the reliefs are set and sized to prevent the vessel from exceeding 1.0 bar in the worst case, so that the vessel is not considered to fall under the scope of the ASME pressure vessel code (At the time of writing this analysis, this was the case for industrially supplied dewars and the laboratory's interpretation of the scope of the ASME pressure vessel code). The vendor does not provide relief valve sizing rationale for the commercially provided dewar, so we will use the above assumptions and back-calculate a flow rate.

The 500 L helium dewars delivered to the laboratory room have two relief valves, one set at 0.55 bar (8 psig) and one at 0.69 bar (10 psig). From the valve literature, the valve cracking 0.55 bar will relieve 4.84 standard cubic meters/minute (std m^3/min) helium (60 SCFM air capacity), at 0.97 bar (14 psig). The valve cracking at 0.69 bar will relieve 3.22 std m^3/min helium (40 SCFM air capacity). With a loss of vacuum, the large flow rate of helium will result in temperatures much lower than room temperature, close to 5 K. A helium temperature of 20 K is conservatively warm and also enough in the ideal gas range for helium that the flow calculation for gas is still accurate. Scaling the flow with the square root of density from 273 to 20 K gives flow rates of 17.9 std m^3/min helium for the 0.55 bar relief and 11.9 std m^3/min helium for the 0.69 bar relief. The total flow rate is 29.8 std m^3/min helium out of the reliefs with 0.97 bar dewar pressure.

Checking other venting scenarios, we conclude that the dewar loss of vacuum drives the highest helium flow rate into the room. Other leaks provide less flow, having lower driving pressure and/or smaller openings. So the case of gas venting from the reliefs, 29.8 std m^3/min helium, is the greatest flow. This is 88 grams/sec or 42.3 L/min, so a full dewar empties in about 12 min.

One full room volume air change per hour would be 520 m^3/h (307 CFM). Our experience tells us that the helium will not significantly mix with the air but will rise to the top of the room. Venting the volume flow rate of helium requires venting 29.8 std m^3/min $\times$ 60 min/h = 1788 m^3/h, which is 3.44 air changes per hour. So either a minimum of 1788 m^3/h (1054 CFM, 3.44 air changes per hour) venting from the top of the room should be continuously provided or this amount of venting should be triggered by the ODH sensor.

Nitrogen ODH Considerations

By far the largest single nitrogen volume which will be brought into the room is that of the 150 L storage dewar, so consider a release of the entire contents of a 150 L LN2 dewar. 150 L warmed to room temperature at one atmosphere would result in 104.5 std m^3 of nitrogen gas. If this were slightly cool and covered the 124 m^2 floor, it would do so to a depth of 0.84 m. If it completely mixed with the remaining air in the room the new oxygen concentration in the room would be 15% or 115 mm Hg. This has a fatality factor of 1×10^{-5}/event according to Fig. 1 of reference [1]. We will use this as the fatality factor for a release of LN2.

Table 4.7 Small laboratory nitrogen ODH analysis summary

Device or action	Failure probability	Number of devices or actions	Total release probability	Fatality factor for LN2 full release	Fatality rate based on release probability
Dewar loss of vacuum	1×10^{-6}/h	2	2×10^{-6}/h	1×10^{-5}/event	2.0×10^{-11}/h
Line leak or rupture	5×10^{-7}/h	2	1×10^{-6}/h	1×10^{-5}/event	1.0×10^{-11}/h
Nitrogen release during transfer	1×10^{-3} per demand	2 × 81/year	1.8×10^{-5}/h	1×10^{-5}/event	1.8×10^{-10}/h
Total fatality rate					2.1×10^{-10}/h
ODH rating					ODH 0

From reference [9], we have the following relevant probabilities:

- dewar loss of vacuum $1 \times 10^{-6}/\text{h}$
- cryogenic fluid line leak or rupture $5 \times 10^{-7}/\text{h}$

There are no U-tube changes for LN2, but transfers are made through foam-insulated tubing. We will treat those transfers as if they were being done through a U-tube, although the operation is safer than that, in order to get an estimate for the probability of a major release. So suppose the probability of a major release is 1×10^{-3}/demand. Laboratory workers report 81 transfers per year, which, taking this to be a typical year, and saying a failure can occur during connect or disconnect, as for helium, gives $2 \times 81 \times 10^{-3}$/year = 1.8×10^{-5}/h.

As for the helium case, there are at most two dewars at a time in the engineering lab. The total probability of a dewar leak or rupture is therefore 2×10^{-6}/h. A line from each dewar to an experimental apparatus would result in at most two lines which could leak or rupture, for a probability of $2 \times 5 \times 10^{-7}$/h that an LN2 line ruptures or has a major leak.

The total probability of a major release of nitrogen is then $(18 + 2 + 1) \times 10^{-6}$/h $= 2.1 \times 10^{-5}$/h. The probability of a fatality due to a release of LN2 is 2.1×10^{-5}/h × 10^{-5}/event = 2.1×10^{-10}/h. This is much less than 10^{-7}, resulting in ODH 0 for nitrogen. The above analysis is summarized in Table 4.7.

Conclusions

Either a minimum of 1788 m^3/h (3.44 air changes per hour) venting from the top of the room to prevent helium accumulation should be continuously provided or this amount of venting should be triggered by an ODH sensor in order to have an ODH 0 rating in the laboratory room. The small amounts of liquid nitrogen present do not present a significant hazard.

References

1. *NASA System Failure Case Studies*, vol. 5, no. 9 (2011)
2. *Buyers Guide* (Cryogenic Society of America, 2017), https://csabg.org
3. D. Arenius, D. Curry, A. Hutton, K. Mahoney, S. Prior, H. Robertson, Investigation of personal and fixed head oxygen deficiency hazard monitor performance for helium gas. Adv. Cryo. Engr. **47**, AIP (2002)
4. M. Mansouri, K.S. Yong, *Technical Description of the Accelerator ODH Detection System—Phase 1*, ESS-0192515 (2018)
5. L. Dufay-Chanat et al., Final report on the controlled cold helium spill test in the LHC tunnel at CERN. Adv. Cryo. Engr. (IOP Conference Series: Materials Science and Engineering, vol. 101, 2015)
6. *ESS Guideline for Oxygen Deficiency Hazard (ODH)*, ESS-0038692 (2016)
7. Oxygen Deficiency SG-M-4-02 CERN (in preparation)
8. *SLAC ES&H Manual Chapter 36: Cryogenic and Oxygen Deficiency Hazard Safety*, http://www-group.slac.stanford.edu/esh/hazardous_substances/cryogenic/
9. Fermilab's ES&H Manual Chapter 4240—Oxygen Deficiency Hazards (ODH), http://esh-docdb.fnal.gov/cgi-bin/ShowDocument?docid=387. Accessed April 2018
10. *Cryogenics Safety Manual Appendix 3: Oxygen Deficiency Hazards*, ANL Physics Division, http://www.phy.anl.gov/division/esh/Cryogenic/Appendix%203/Appendix%203.htm
11. *Oxygen Deficiency Hazard Control (ODH) Control Program*, Jefferson Lab ES&H Manual, https://www.jlab.org/ehs/ehsmanual/6540.htm
12. M. Chorowski et al., Safety oriented analysis of cold helium-air mixture formation and stratification. Cryogenics **46** (2006)
13. M. Vadon, B.P. Solano, F. Baida, Simulations of liquid argon accidents in the ATLAS tunnel. Adv. Cryo. Engr., **47**, AIP (2002)
14. W.M. Soyars, J.L. Schiller, Open channel helium flow during rupture event. Adv. Cryo. Engr. **47**, AIP (2002)
15. J. Fydrych, E. Lundh, J. Moberg, D. Phan, P. Tereszkowski, J.G. Weisend II, The safety helium collectors for the ESS superconducting linac: functional specification and detailed design, *Presented at 2018 International Cryogenic Engineering Conference* (at press)
16. E. Lundh, *Results from CFD simulations—Helium spill in the tunnel*, ESS-0061985 (2016)

Chapter 5
Oxygen Safety

John M. Jurns

SYLPHS! From each sun-bright leaf, that twinkling shakes, 'Oer Earth's green lap, or shoots amid her lakes, your playful bands with simpering lips invite, and wed the enamour'd OXYGENE to LIGHT.

—Erasmus Darwin, "The Botanic Garden A Poem in Two Parts" 1791.

Abstract This chapter begins with an informative example of an accident involving oxygen, emphasizing the importance of understanding the safe use of oxygen. A short history of oxygen—its discovery, development as a commodity, production and distribution is also included to provide some background to current use and safety issues. The rest of the chapter is devoted to laying a foundation for the reader to obtain an understanding of how to safely work with oxygen and oxygen systems. We start with an overview of oxygens properties—physical, optical, magnetic, and oxygens allotropes. Following is a section on basic principles for the safe use of oxygen. This includes safety issues related to both liquid oxygen (LOX) and gaseous oxygen (GOX). One of the most important issues concerning oxygen safety is oxygen concentration—oxygen enrichment (too much oxygen), and oxygen deficiency (too little oxygen). Following is a section on personnel safety and health, elaborating on appropriate precautions, training, practice, and emergency response. A section on system design is provided to give the reader guidance on good rules to follow when designing oxygen systems—steps to follow during design, compatibility of materials with oxygen, facility planning, manufacturing, installation and operations. The chapter concludes with a section on best practices and a list of references useful for anyone designing, building, or operating oxygen systems.

T. J. Peterson and J. G. Weisend II, *Cryogenic Safety*, International Cryogenics Monograph Series, https://doi.org/10.1007/978-3-030-16508-6_5

5.1 Example Accident

This incident occurred back in the early 1970s at a NASA launch facility. It occurred at a time that the Apollo missions were coming to an close, but NASA still had a robust program of launches. That is to say, they had several decades of working successfully with large quantities of cryogenic oxygen and fuel.

Before the start of this incident, security guards had driven their security cars to a launch pad during clearing operations to inspect the area. The drove to the area, parked their cars, shut off the engines and got out to conduct their inspection.

The launch pad area contained a large spherical liquid oxygen (LOX) dewar. As the guards were conducting their inspection, LOX was also being dumped into a drainage ditch as part of normal operations. The LOX vaporized and created an oxygen cloud, which drifted into the path of the cars. The amount of LOX dumped was approximately 40,000 L, sufficient for an oxygen cloud to cover 20 acres (81,000 m^2) a depth of 1.2 m at 40% oxygen concentration. The engine compartments in the vehicles caught fire by auto-ignition from engine heat, combustibles (either oil on the engine or possibly vaporizing gas from the carburetor), and the enriched oxygen atmosphere. The results can be seen in Fig. 5.1.

There are several lessons that can be learned from this incident. The first observation is that personnel were present during the LOX dumping operation. Unfortunately, the report detailing this incident [NASA TM X-67953] did not give any additional details, but it was probable that this was all done according to procedure, the risks evaluated, and the decision made to allow personnel to be present during dumping

Fig. 5.1 A very bad day at work (reproduced from [2])

LOX. One lesson learned would be to avoid having personnel near where LOX was being dumped.

A second lesson has to do with the very act of dumping liquid oxygen. When one spills or dumps cryogenic fluids, even after they vaporize they are denser than air until they warm up. This means that you can have exactly what happened here—a cloud of air with high oxygen concentration that can persist for an appreciable period of time. If liquid oxygen needs to be drained from a dewar or piping system, it is always better to vaporize it first instead of dumping liquid. This reduces the risk of an atmosphere with high oxygen concentration.

A final lesson has to do with personal protective equipment (PPE). A portable oxygen monitor could have alerted personnel that there was a high concentration of oxygen, and they could have taken appropriate action to protect themselves.

This event occurred over 40 years ago, but oxygen (both gas and cryogenic liquid) is still used extensively today in many different areas. One thing that is important to remember when dealing with oxygen systems and safety is that oxygen has many more *unsophisticated* users than pretty much any other cryogenic fluid. There are people from all walks of life that are using oxygen that are only vaguely aware of the risks. It is incumbent on the professional to do all they can to ensure that oxygen systems are as safe and foolproof as possible.

5.2 History

5.2.1 Discovery

Like many other scientific discoveries, more than one person was making investigations that lead to the discovery of oxygen. Carl Wilhelm Scheele (Sweden) in 1772 (published 1777), Joseph Priestley (England) in 1774 (published 1775) and Antoine Lavoisier (France) in 1774 (published 1777) all contributed to the discovery of oxygen. Although we consider it an obvious fact today, they discovered that oxygen is required for respiration and for combustion. Priestly and Scheele both produced oxygen as a result of heating metal oxides (Priestly HgO, Scheele MnO_2), but didn't necessarily understand that it was a component of air. However, Priestly did discover that his "dephlogisticated air" supported both respiration and combustion "five or six times as good a common air". Scheele also discovered many other chemical elements in his research, but unfortunately died rather early in life from the cumulative effects of many of the toxic substances he worked with. Apparently he had the bad habit of smelling and tasting any new substance he discovered. Lavoisier correctly identified oxygen as one of two components of air (the other being *azote* or nitrogen). He also identified the correct role of oxygen in the process of combustion. However, it wasn't until John Dalton had developed his atomic theory decades later that oxygen was correctly understood to be a unique element.

Priestly

Scheele

Lavoisier

5.2.2 *Production and Distribution*

Fast-forward about a hundred years, and in France, we find Louis Paul Cailletet producing liquid oxygen by cooling highly compressed oxygen and expanding it through a Joule-Thomson device to produce droplets of liquid. A commercially viable process for producing oxygen was developed soon afterward by both Carl Von Linde (Germany) and William Hampson (England) in 1895. They liquefied air by reducing its temperature, and then separated the oxygen and nitrogen by fractional distillation.

Modern production still uses fractional distillation of air to separate it into its components (mainly oxygen, nitrogen, and argon). Pressure swing absorption is another method of producing oxygen by passing air through a bed of zeolite molecular sieve that preferentially adsorbs nitrogen and produces an oxygen stream of 90–93% purity. Membrane separation technology is yet another method of producing oxygen by passing air through an engineered membrane filter that selectively separates oxygen from other gasses. Another method of producing oxygen is by the decomposition of water by electrolysis into its hydrogen and oxygen components. Other less common chemical methods have also been used to produce oxygen.

High purity oxygen is typically distributed to the market either in high-pressure gas cylinders at room temperature, in portable cryogenic liquid dewars, or in cryogenic liquid trailers. Liquid containers have the advantage of being able to transport much larger quantities of product due to liquid oxygen's higher density. However, high-pressure gas cylinders have the advantage of transporting oxygen without product loss.

5.2.3 Accidents

Although oxygen and it's role in combustion (oxidation) was discovered over two hundred years ago and has been in production well over one hundred years, the hazards of working with pure oxygen were unfortunately identified largely through a long history of accidents, trial, and error. A survey of oxygen related accidents shows that not much was reported in the early days of oxygen delivery. Much of the data available is from government sources, and shows a preponderance of accidents related to both oxygen therapy and oxy-acetylene welding/cutting. No wonder, as in the United States alone, there are approximately one million patients receiving long term oxygen therapy [1], and oxy-acetylene torches are ubiquitous fixtures in shops and manufacturing facilities.

Accidents in oxygen systems typically fall into the following categories:

- Material incompatibility
- Materials failure
- Design deficiency
- Cleaning deficiency
- Procedural deficiency

Figure 5.2 shows the relative percentage of accidents due to these various causes from an investigation made by NASA in 1971 [2]. Although this chart is specific to NASA mishaps, it is reasonable to draw a general conclusion that a majority of accidents in oxygen systems can be attributed to either procedural or design deficiencies.

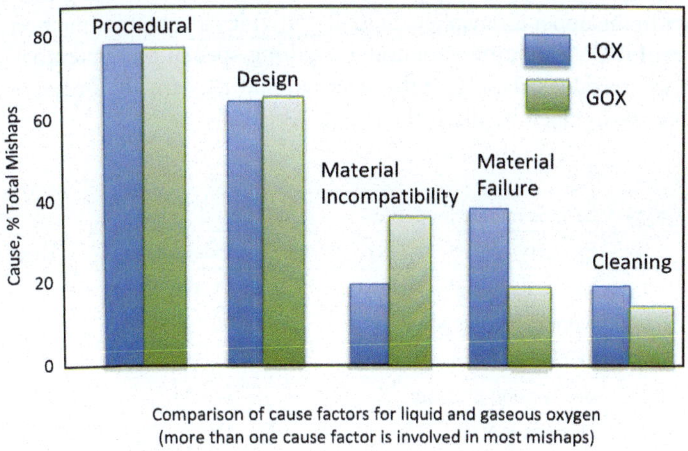

Fig. 5.2 Cause factors of accidents in oxygen systems (reproduced from [2])

5.3 Properties

Physical properties—Oxygen can exist as a solid, liquid, or gas. Liquid oxygen is pale blue color. Oxygen gas is odorless, colorless and tasteless. Oxygen solidifies at atmospheric pressure and 54.36 K, and is also a light blue color. However, solid oxygen contains a number of crystalline phases depending on temperature and pressure [3] (Table 5.1).

Enthalpy, entropy, viscosity, thermal conductivity, surface tension, etc. are all properties that depend on both the temperature and pressure. Values can be found at the NIST thermophysical properties website [4] or from the commercially available fluid properties database REFPROP [5].

Optical properties—Oxygen liquid is a light blue color, and has an optical refractive index of: 1.000271 (gas) and 1.221 (liquid).

Magnetic properties—LOX is slightly magnetic (the only atmospheric gas with this property). Paramagnetic susceptibility is 1.003 at NBP [6].

Allotropes—O, O_2, O_3. Atomic oxygen (O) exists as a single atom of oxygen that quickly bonds with other molecules. It is highly reactive. It does not exist in any quantity normally near the Earth's surface, but it is the primary species in near space (low earth orbit), formed by UV radiation of oxygen. In low earth orbit, atomic oxygen's propensity to combine with other molecules makes it a significant problem for spacecraft [16]. Diatomic oxygen (O_2) is the most common form of elemental oxygen found on Earth. It forms 20.9% of the Earth's atmosphere by volume. Ozone (O_3) is also a reactive form of oxygen that exists in the atmosphere. It is produced by various means, the most familiar being the result of an electrical discharge in air. Although considered a pollutant, and harmful to people with lung conditions, the ozone layer in the upper atmosphere actually provides a valuable function as a shield against harmful UV radiation. Not doing anything special on September 16th? You may want to consider celebrating the International Day for the Preservation of the Ozone Layer designated by the UN General Assembly.

Table 5.1 Oxygen thermodynamic properties

Symbol	Property	Value	Unit
MW	Molecular weight	31.999	–
T_{nbp}	Normal boiling point	90.19	K
T_{tp}	Triple point temperature	54.36	K
P_{tp}	Triple point pressure	146.33	Pa
ρ_{NTP}	Density at NTP	1.327	kg/m^3
ρ_{nbp}	Normal boiling point density (gas)	4.467	kg/m^3
ρ_{nbp}	Normal boiling point density (liquid)	1141.2	kg/m^3
k	Specific heat ratio	1.4	–
duh_{fg}	Latent heat of vaporization (nbp)	213.06	kJ/kg

5.4 Basic Principles for the Safe Use of Oxygen

Oxygen itself is not flammable. However, it is a necessary ingredient in combustion, and will cause materials that normally burn to combust at an accelerated rate. Consequently, care must be taken when working with oxygen to prevent high concentrations, specifically around potentially combustible materials. The "fire triangle" (Fig. 5.3) provides a good graphic to show that oxygen, fuel, and heat are all required for combustion. If any one of these elements is missing, combustion is impossible.

5.4.1 Liquid Oxygen Safety

Standard cryogenic safety—Cryogenic fluids fall into three categories: inert gases, flammable gases, and oxygen. Common inert gases include helium, nitrogen, and argon. These gases are typically used as purge gases. They are not flammable and do not support combustion. Flammable gases include (but are not limited to) gases such as hydrogen or methane. These gases can burn or explode under the right conditions given an ignition source and oxygen. As mentioned previously, oxygen is not flammable in itself, but will support combustion. All cryogenic fluids have hazards associated with them, but are safe if used correctly. Hazards that are common to all cryogenic fluids basically fall into three categories:

- Cryogenic burns, hypothermia or frostbite from contact with cold fluids or surfaces (see also Chap. 2)
- Over pressure and possible explosion from trapped cryogens vaporizing and expanding (see also Chap. 3)
- Atmospheric hazards—either oxygen deficiency or oxygen concentration (see also Chap. 4).

Cryogenic burns result as either from direct contact with liquids, direct contact with cold venting gas, or contact with cold equipment in cryogenic service. Cryogenic burns can cause frostbite—freezing of human tissue. Frostbite is evidenced by a change in color of skin to white or gray, possibly followed by blistering. Deep frostbite is indicated by a generally waxy appearance to skin. Care must also be

Fig. 5.3 The fire triangle

taken to prevent cryogenic liquids from dripping on clothing. As cryogens have viscosities anywhere from 1 to 23% of water, they are more prone to being absorbed and retained by clothing. Besides risk of frostbite, cold venting liquid or gas directly to the face and eyes can result in blindness. In all cases, it is critical to have appropriate Personal Protective Equipment (PPE) when handling cryogenic equipment to prevent unnecessary injuries. PPE will be discussed later in this chapter.

Over pressure in a vessel or piping can occur when a cryogenic fluid is trapped with no means of venting, for example between two isolation valves in a pipe. The trapped cryogen can then vaporize from ambient heat and dramatically increase in pressure. This can then result in a BLEVE—boiling liquid vapor expansion—explosion. To prevent this, safety relief valves should always be installed between isolation valves in cryogenic piping, and on pressure vessels, as shown in Fig. 5.4.

Another form of potential cryogen trapping is an ice plug. Ice plugs can form when moisture from the air condenses inside a cryogenic line. A frozen ice plug can form, effectively blocking the line and preventing vaporizing cryogen from venting. The risk of an ice plug increases as the diameter of the vent line decreases. Frozen plugs can form in dewar plumbing if the cryogenic system is exposed to air in relief devices or vent lines. Moisture in air can block line. The smaller the vent line, the more susceptible it is to plugging. Air can be prevented from coming in contact with cold piping or valves by the use of check valves, and by purging lines with an inert gas.

Ice build up inside a pipe is not the only risk arising from moisture condensation. Moisture can condense and freeze into a block of ice on the outside of piping and valves. If the ice ball becomes large enough, it can block the outlet of relief valves, prevent the function of valve operators, or freeze non-cryogenic rated components such as elastomeric seals. Diligence is required to maintain cryogenic systems where there is a risk of ice formation to assure that ice does not interfere with the operation of equipment.

Cryogenic safety specific to oxygen—liquid oxygen specific hazards relate largely to oxygen concentration in the air. Normally, oxygen constitutes approximately 21% by volume of ambient air. Cold oxygen vapor is denser than air, and venting or leaking oxygen can result in higher concentrations of oxygen. Vent stacks for oxygen containers should be considered to avoid venting pure oxygen in an area. A good reference for oxygen vent systems is found in EIGA 154/16 [7]. Oxygen enrichment can occur in cryogenic systems using helium or hydrogen that are not properly insulated. Un-insulated cryogenic helium and hydrogen can liquefy ambient air. As oxygen liquefies at a warmer temperature than nitrogen, oxygen will preferentially liquefy first.

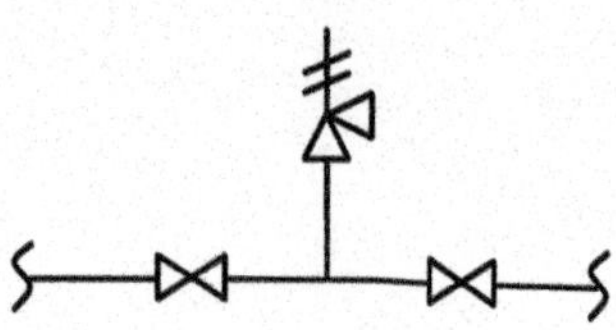

Fig. 5.4 Safety relief valve installed between two isolation valves

5.4.2 *Gaseous Oxygen Safety*

As oxygen gas is invisible and odorless, it is not readily detectible. High or low concentration of oxygen in the atmosphere poses specific hazards, and steps must be taken to mitigate these hazards. Considerations such as oxygen deficiency monitors, proper ventilation, and placement of equipment must be included in design. High-pressure oxygen gas poses significant risk. As with all pressure equipment, proper design for pressure containment, safety relief and guards must be made to assure a safe and reliable system. Pressure safety is discussed in detail in Chap. 3.

5.5 *Oxygen Enrichment*—What Happens if There is Too Much Oxygen?

5.5.1 *Causes*

Oxygen enrichment occurs when leaks from oxygen sources find their way into the ambient atmosphere. Leaks can come from sources such as liquid dewars, high pressure cylinders, valves, or piping. Care must be taken to install and maintain equipment properly to minimize leaks from gaskets, seals, and valve leaks. Additionally, the concentration of oxygen in the atmosphere can increase from un-insulated cryogenic surfaces. Cryogenic nitrogen and hydrogen systems operate below the liquefaction temperature of atmospheric oxygen, meaning that oxygen can preferentially condense on any un-insulated surfaces, resulting in increased local oxygen concentration.

Hazards

The hazards associated with oxygen enrichment fall into the categories of *fire* and *health*.

Higher oxygen concentration increases the risk of *fire*. Materials generally ignite at a lower temperature with higher oxygen concentrations. Materials that will burn in a normal atmosphere (~21% oxygen) can burn much more vigorously where the oxygen concentration is increased. Some materials will ignite with only a shock impact (such as a hammer blow or falling piece of equipment) when exposed to liquid oxygen.

Personnel working with oxygen need to be aware that clothing can absorb oxygen, and the risk of high oxygen concentration can persist even after they have moved away from an area where high oxygen concentration is present. Exposure to open flames can result in clothing catching on fire and burning vigorously, possibly resulting in significant personal injury.

Fires can spread rapidly in oxygen-enriched atmospheres, and automatic detection and extinguishing systems are recommended. Detailed recommendations on fire extinguishing systems can be found in NFPA 53—"Recommended Practice on Materials, Equipment, and Systems used in Oxygen Enriched Atmospheres" [8].

Table 5.2 Effects of hyperoxia

Effect	Consequences	Potential result
Chemical toxicity Capillary endothelium Alveolar epithelium	Pulmonary damage Atelectasis	Hypoxemia Acidosis Death
Retinal damage	–	Blindness
Toxic effects upon enzymes and cells of the central nervous system	Twitching Convulsions Destruction of neurons Death	Death
Erythrocyte hemolysis Myocardial damage Renal damage Hepatic effects Chemical toxicity Destruction of cells	Endocrine effects—adrenal, gonads, thyroid	Death

Besides risk of personal injury from fire, there are also other significant *health risks* associated with oxygen-enriched atmospheres. Oxygen toxicity or hyperoxia can occur with prolonged exposure to oxygen concentrations greater than 50%. Physiological health problems can be manifested in different areas such as (for example) eyesight, kidneys, the pulmonary system, or the endocrine system. As shown in Table 5.2, consequences of hyperoxia can be as serious as death [9].

5.6 *Oxygen Deficiency*—What Happens if There is Too Little Oxygen?

5.6.1 *Causes*

Perhaps a more common problem is that of oxygen deficiency. As with the previously mentioned issue of high oxygen concentration, leaks from piping, valves and seals can occur, and leaks coming from other gas sources can displace oxygen in the air. A decrease in oxygen concentration in the air is a serious problem, and is exacerbated by the fact that a low oxygen concentration is not detectable if oxygen in the air is displaced by an inert gas. Spills or leaks of inert gases such as nitrogen, helium, or argon can reduce the normal concentration of oxygen below survivable levels. This risk is particularly high when working with cryogenic liquids. Vaporizing cryogens have a vapor to liquid ratio as high as 900:1, meaning that a small amount of vaporizing liquid can readily displace the normal atmosphere, resulting in low oxygen concentration levels.

Table 5.3 Health risk from oxygen deficiency [14]

19.5%	Minimum acceptable oxygen level
15–19%	Decreased ability to work strenuously, impaired coordination
12–14%	Respiration increases, poor judgment
10–12%	Respiration increases, lips blue
8–10%	Mental failure, fainting, nausea, vomiting, unconsciousness
6–8%	8 min-fatal 6 min–50% fatal 4–5 min-possible recovery
4–6%	Coma in 40 s, death

5.6.2 *Hazards*

The primary health hazard from oxygen deficiency is hypoxia. Hypoxia can occur when oxygen concentration falls below 19.5%. Symptoms can include the following:

- Lightheadedness/fatigue
- Numbness/tingling of extremities
- Nausea
- Ataxia (heart palpitations)
- Confusion/disorientation
- Hallucinations/behavioral change
- Breathlessness
- Heart malfunction
- Death

Table 5.3 shows the symptoms associated with oxygen deficiency, and the progressive increased health risk as the oxygen concentration decreases.

Additional details on oxygen deficiency hazards and mitigations may be found in Chap. 4.

5.7 Personnel Safety and Health

When working with oxygen systems, adequate precautions should always be taken to minimize the risk of personnel injury. These precautions include (but are not limited to) proper ventilation, cleanliness of equipment and the work environment, routine inspection of equipment, adequate personnel training, and personal protective equipment (PPE).

Proper ventilation minimizes the risk of either high or low oxygen concentration, both of which have their unique hazards. Work areas should be free of combustible material, and dirty (oily or greasy) tools should not be used when working with oxygen. Inspecting equipment used in oxygen service is critical, especially equipment that is moved around a lot. Equipment such as oxy-acetylene torch kits or medical oxygen concentrators and gas cylinders are subject to greater wear and tear than more static equipment. It is also possible that they could be moved into areas that are not as clean as ideally should be, requiring additional caution when using.

Personal protective equipment is a key element of maintaining personal safety. The appropriate equipment should be chosen based on the particular hazards encountered. When working with liquid oxygen, appropriate PPE includes a face shield to protect from splashes of cryogenic liquid, insulated gloves rated for cryogenic service to prevent contact of bare skin with cold surfaces, Choice of clothing is also important. When working with cryogens, it is best to avoid clothing with open pockets or cuffs that could trap spilled liquid. Additionally, clothing that is "flame resistant" or "flame retardant" may still be flammable in a high oxygen concentration environment. When working with open containers of liquid oxygen, an added precaution of an impermeable apron to protect from splashes is recommended. Long pants should be worn, with the pant legs outside the shoes. Shoes themselves should not be a permeable material such as canvas that could soak up spilled cryogens. When working in an area where oxygen concentration could be either high or low, it is also advisable to wear an approved portable oxygen monitor to provide warning is also advised.

Another critical element of maintaining personal safety is training. Personnel in contact with oxygen equipment must understand what they are working with, and the hazards and risks associated with it. Organizations working with oxygen systems should institute a defined program of training for personnel working with these systems. Appropriate training for working with oxygen systems includes:

Knowledge—Training should include information on gaseous and liquid oxygen fluid properties, compatibility of materials with oxygen, safe procedures for working with and maintaining oxygen equipment, hazards associated with oxygen systems, confined space training, correct use of PPE, and emergency response.

Practice—Although *knowledge* is important, just as important is *practice*. Organizations working with oxygen systems should institute a program to maintain worker competency. This would include practice with procedures for operating equipment, filling and venting, transportation, equipment maintenance and rebuilding, and testing (such as pressure and leak testing). As equipment becomes obsolete and is replaced by new equipment, training must be kept up to date to verify that personnel competencies are current.

5.7.1 Review/Maintenance

Organizations using oxygen systems should institute a regular program of review and maintenance. Manufacturers of oxygen equipment typically include instructions that stipulate service schedules, lifetime for rotating equipment, and soft goods such as seals to be replaced. Record keeping is a vital aspect of this element—diligently logging schedule and maintenance performed on equipment helps assure that equipment is in good working order, and that items have not been overlooked. Records should include at a minimum the date the equipment was put into service, schedule time for maintenance, log of current condition, and what parts were replaced.

5.7.2 *Certification*

Depending on the type of service oxygen equipment is put into, different certifying bodies will have jurisdiction. For example, pressure vessels may be subject to the ASME Boiler and Pressure Vessel code, or the European Pressure Equipment Directive (PED); equipment transported over the road may be subject to the U.S. Department of Transportation (DOT) regulations. Pressure piping may be subject to either ANSI or EN regulations. High-pressure gas cylinders would typically meet 49 CFR or PED requirements. It is up to the responsible engineer to determine what codes apply to their equipment and arrange for the equipment to be certified according to applicable regulations.

5.7.3 *Emergency Response*

In emergency situations, it is always most important to protect personnel. Trained personnel, who have the proper equipment to deal with these situations, should handle emergency response. Oxygen related emergencies typically are from overpressure, leakage or spillage of gaseous or liquid oxygen in a system. This could be due to impact on the equipment, tipping or dropping equipment, especially in the case of equipment being transported, or improper operation. The result could be fire, cryogenic burns, explosion with resulting shrapnel, or crushing.

As personnel safety is of primary importance, first aid should be administered if required. In the case of oxygen deficient atmosphere, it is important to remember that if personnel lose consciousness, *do not attempt to remove them* without first determining that the atmosphere can be safely entered, or SCBA (Self Contained Breathing Apparatus) is available (and can be used by trained personnel). In the case of cryogenic burns, appropriate first aid should be given. This is covered in Sect. 2.5.

In the case of fire, again, trained personnel should respond. Fires in oxygen rich atmospheres burn more vigorously than in a normal atmosphere. If possible, the source of oxygen should be isolated. Depending on the size and strength of the fire, it may be better to try and contain the fire and let it burn out. NFPA 53 provides direction on fire fighting for oxygen rich fires.

5.7.4 *Buddy System*

When working with oxygen equipment, or pressurized and cryogenic systems in general, it is best to use a "buddy" system. That is, no individual should be working in a hazardous situation alone, but if one person is operating equipment, a second person should be present to observe and be able to react to any emergency situations.

5.8 System Design

5.8.1 *Overall Guidelines*

When designing systems for use with oxygen, it is important that the individuals performing the engineering design are qualified to perform the work. This does not necessarily mean having a "oxygen certified engineer" rating; indeed there is no standard in the industry for this. However, designers should be familiar with the unique properties and hazards associated with oxygen, and have experience designing similar systems. Additionally, personnel that are responsible to review these designs should also be competent in the design and use of oxygen systems.

Designs for oxygen systems must be in compliance with applicable standards. It is the responsibility of the designer to determine what standards use, based on the application. For example, equipment designed for over-the-road transport must be designed in accordance with the Code of Federal Regulations (49 CFR in the United States). There are many standards that specify how oxygen equipment and system should be designed, built, installed, and maintained. Some are government national or international standards, and others industry standards. A list of standards is included in the references at the end of this chapter. Some organizations with oxygen standards include:

- NFPA—The National Fire Prevention Association provides guidance on installation of oxygen systems, where they should be located, distances from other structures, barriers, controls, fire suppression, and related issues
- CGA—The Compressed Gas Association has many standards, relating to sizing safety relief devices, standards for piping and vent systems, standards for valves and fittings, and installations
- EIGA, AIGA—The European Industrial Gas Association and Asian Industrial Gas Association also are a valuable repository of information on oxygen systems, and parallel many of the standards maintained by the CGA
- ASTM International—ASTM also has many standards; those relating to oxygen deal with standards for testing, leak detection, and cleaning
- CFR—The Code of Federal Regulations is a huge repository of regulations on almost everything. 49 CFR deals largely with design and rules for transporting oxygen

- Local and organization specific regulations—Besides national and international standards, the designer must also check to see what local regulations may govern for oxygen systems they design, build and install.

When designing oxygen systems, one needs to consider factors such as:

- System dynamics—is this a stationary or moveable system? Are there large changes in temperature, pressure or flow? What materials are involved—will any flammable material be exposed to oxygen? What are the operational constraints? Specifically, who are the users—are they trained professionals or unsophisticated operators? What is the environment that the equipment or system will be used? Are there adequate barriers and protections for personnel? Is the equipment or system to be operated inside or outside? That is—adequate protection from weather or other conditions is required.
- Use oxygen compatible materials as much as possible. This will be discussed in detail later in the chapter. The designer needs to be aware of issues related to materials and oxygen compatibility. This is not a straightforward issue, and engineering judgement and analysis is required to determine which materials are suitable for the design.
- Particulates. These are particularly nasty problems in oxygen system, as they can be ignition sources in piping, or contamination damaging equipment. A good practice is to use filters to capture particulates (and make them removable for inspection)

Designing system and equipment for oxygen service should follow the same principles as any good design. Once the requirements for the system have been defined, one starts with a *design analysis*, followed by *manufacturing and fabrication* using safe and efficient processes. Once equipment or systems have been built, it is important to perform appropriate *tests* to validate their function. These could be as simple as pressure and leak tests for pressure vessels and piping, or as complicated as detailed functional tests to prove that the equipment performs as designed. After equipment has been designed, built, and tested, it must be *commissioned* at its final location. Finally, a *maintenance plan* should be implemented to make sure the equipment continues to operate effectively over its entire life cycle.

Risk Training

Design and operation of oxygen systems requires that all personnel associated with it understand risks associated with oxygen systems. This includes the entire chain of people from management, to engineers designing the equipment, to technical personnel operating equipment, and finally to customers using either oxygen or oxygen generating equipment. A preponderance of oxygen related accidents come from end users that are either not adequately trained, or choose to ignore the hazards and risks associated with oxygen.

More detailed summaries of this topic is found in Sects. 2.10.1 and 8.3. The basic elements to be included in any risk training are as follows:

- Identify the hazards—We point out here that a *hazard* is not the same as a *risk*. A *hazard* is something that has the potential to harm—events such as oxygen deficiency, cryogenic burns, or other dangers inherent in a system. A *risk* is the probability that a hazard will result in damage or injury. It may be possible to design a hazard out of a system, but in many cases this is not possible, and it is the responsibility of the designer to implement design or operational features that will *minimize the risk* of a hazard resulting in damage or injury.

- Risk mitigation & control—Once hazards have been identified, their severity must be evaluated, and risk mitigation steps taken to bring the risks to an acceptable level. These mitigation steps may be either changes in the design, or changes to procedures.
- Procedures to follow in case of an accident—In case of an accident, procedures must be put in place on how to deal with the accident. This includes for example notifying first responders, first aid for personnel, how to secure equipment, and what steps to take to restore equipment or service.
- Follow up—Finally, after an incident, it is important to conduct a formal review to determine how to avoid future accidents, assess the equipment or system to determine what needs to be fixed or replaced, and updating designs and procedures.

Specifications

- Meet technical requirement—this may seem obvious, but it is important to make sure that the system or equipment being designed adequately provides the needs specified. Specifications should be clearly stated by the customer or user, and the proposed design checked against those specifications. Diligent work done at this stage will prevent re-work later on in the execution of the design.
- Meet regulatory and code requirements—once the technical requirements have been properly defined, the designer must decide which regulatory and code requirement must be met. For example, pressure vessels and piping above a certain pressure must meet ASME section VIII, ANSI, or PED codes before being put into service. Installations at consumer sites such as hospitals must meet certain quantity/distance requirements such as specified in NFPA regulations.
- Pay attention to material selection—oxygen compatible materials must be used to the greatest extent possible, remembering that what material can be used is dependent on a number of factors such as temperature, pressure, and flow.
- Don't over-specify—the design of any component or system requires consideration of trade-offs between competing factors such as reliability, cost, schedule, etc. When specifying materials for oxygen related equipment and systems, careful attention should be paid to choice of materials. As will be discussed in section titled "Material Compatibility", choosing materials for use in oxygen service is not a cut-and-dried matter, but requires analysis and judgement to determine if the material chosen is appropriate for the intended use. For example, stainless steel piping is an appropriate material to use for cryogenic oxygen service based on it's allowable operating temperature range. However, if the system will only see warm oxygen gas service, in many cases carbon steel piping is entirely adequate.

5.8.2 *Reviews*

- Reviews should be conducted regardless of the size or complexity of the system under consideration. A formal, rigorous process will help to insure a safe system. Reviews should include personnel qualified in applicable fields of expertise, and are independent of the project to maintain objectivity.

5.8.3 *Typical Review Stages*

- Design—at this stage, review the overall design—does it meet the technical and regulatory requirements? Is the design cost effective and buildable? At this stage the fundamental design and operation concept needs to be adequately defined before moving on to detailed design and other reviews.
- Safety/Hazards—this review includes analysis of what the safety hazards are, and what the risks are. Note that a risk is not the same as a hazard—a hazard is something that is dangerous, and a risk is how likely that hazard will result in an unwanted event. After risks are determined, analysis must show what steps must be taken to mitigate them to an acceptable level. Different processes are available to conduct a hazards analysis, and details of how an analysis is performed are outlined in more detail in 2.10.1 and 8.3. The general format of this exercise is to:
  - Identify the hazard
  - Determine the risk and assign a rating to that risk
  - Specify how the risk will be mitigated
  - Determine a final risk rating based on the mitigation strategy
  - Note that a design is not considered acceptable until all identified risks have been mitigated to an acceptable degree
- Operational—this review will examine how the system or component will be used. Even though safety features should be designed into an oxygen system, it still must be operated and maintained in a manner that will not result in injury to the user or damage to the equipment. Detailed procedures should be developed and written out in clear, unambiguous language to allow a user to safely operate the system or component. Particular attention should be paid to writing instructions and procedures appropriate to the level of sophistication and familiarity the user has with oxygen systems.

5.9 Design of Gaseous Oxygen (GOX) Systems

This section will provide some general guidelines in the design of equipment and systems for gaseous oxygen service. The definition we will use for gaseous oxygen service generally pertains to equipment operating at ambient temperature and above, and at pressure above atmospheric pressure.

- Minimize sources of particles—particles in a system flowing oxygen can provide either fuel (depending on the material), or energy (from impact of particles on a surface) that can result in combustion in an oxygen-rich environment. Therefore, it is critical for designs to minimize risk of particles. This is accomplished several ways:
  - System cleanliness—first of all, a system should be cleaned as much as possible to eliminate any particulate matter from fabrication, such as dust, metal chips, weld slag, or other foreign materials.
  - Avoid rotating components—rotating valves or seals can generate particles by shearing surfaces against each other. It is better to provide valves with non-rotating stems.
  - Minimize burrs, sharp edges, and soft material extrusion—burrs and sharp edges can provide an ignition source if a particle impacts them, and also possibly break off in high velocity flow, providing combustible material. Design must also eliminate the potential of soft materials such as O-rings being extruded or sheared, providing potential combustible material.
  - Avoid crevices—fabrication methods that minimize the presence of crevices are preferred, as crevices can be a place where particles accumulate. For this reason, it is always preferable to use butt-welded joints instead of fillet welded or socket welded joints (as shown in Fig. 5.5).
  - Use filters—judicious placement of filters in oxygen systems can catch any particles that do find their way into the system, and are therefore a common and effective method for minimizing risk of system particulates and consequently particle impacts in the system.
  - Limit gas pressure—a good rule of thumb is to reduce pressure in a system as early as possible. That is, if the system has some length of piping or tubing, it is preferable to reduce the pressure to the final use pressure closer to the source, as combustibility of materials is directly related to the oxygen pressure. Of course this involves trade-offs in design, as a lower pressure system may require greater diameter piping for the same flow rate, thus potentially increasing the cost.
  - Limit gas pressurization rate—too fast a pressurization rate may result in sufficient heat of compression to initiate combustion.
  - Limit gas velocity—Greater velocity in piping means greater energy for any particles that may be entrained in the gas stream. As mentioned in the above paragraph, reducing system pressure will reduce risk of ignition of either particles or the piping itself. Figure 5.6 shows the acceptable relationship between velocity and pressure for carbon steel, stainless steel, and non-ferrous piping.

The curve shown in Fig. 5.6 is valid for temperatures up to 150 °C (302 °F) for carbon steel piping, and 200 °C (392 °F) for stainless steel and non-ferrous piping. The carbon steel temperature limitation may be increased to 200 °C (392 °F) provided a hazard analysis is performed that takes into account factors such as site conditions,

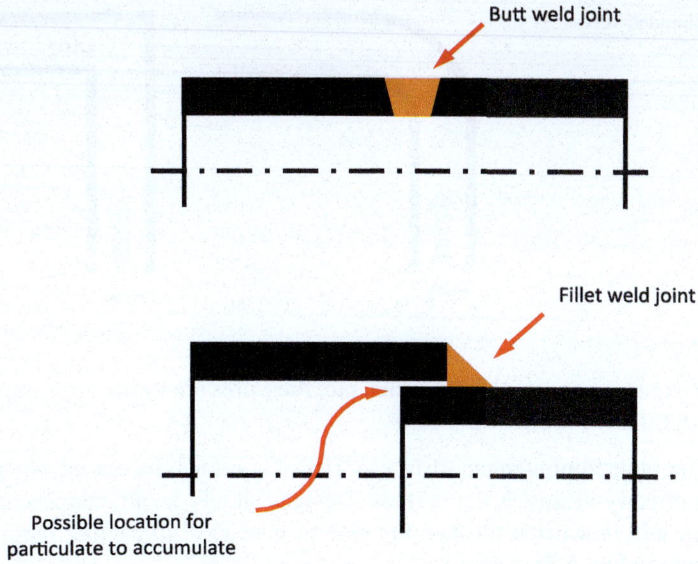

Fig. 5.5 Butt weld versus fillet weld pipe joint

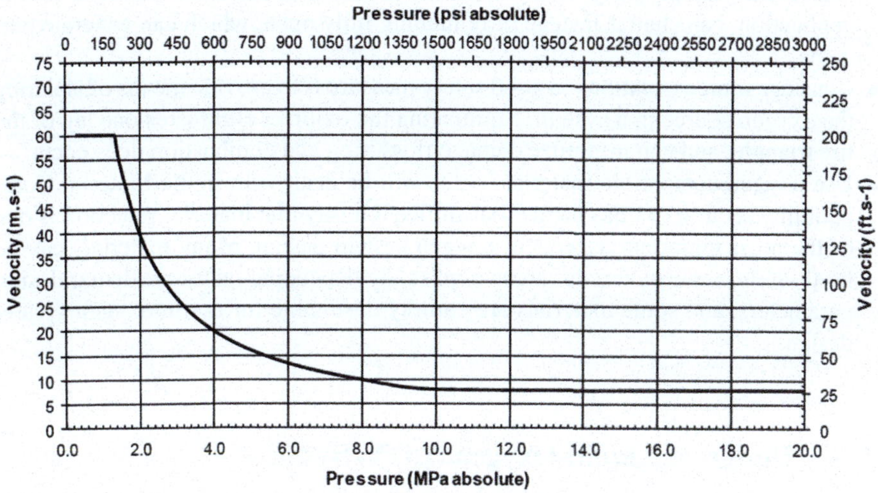

Fig. 5.6 Non-impingement velocity curve for oxygen in piping [10]

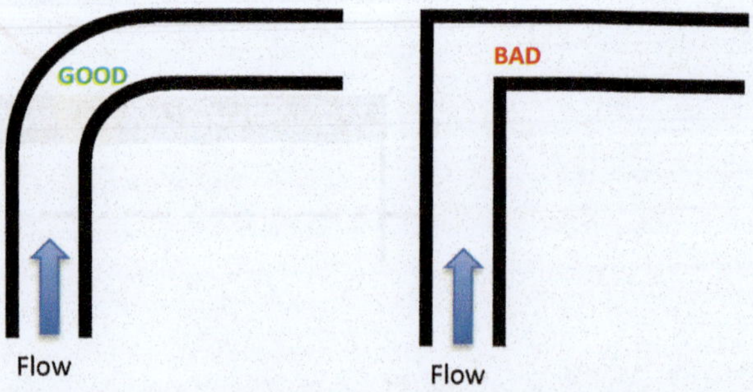

Fig. 5.7 Rounded versus sharp edge flow

operating experience, experimental data, etc. Pressures are limited to a maximum of 21 MPa (3000 psig) [10].

- Minimize blunt impingement surfaces—risk of ignition is increased when particles impact directly against a flat surface. Designs should avoid sharp bends so any particles will be carried through the system instead of impacting piping surface (as shown in Fig. 5.7).
- Avoid component and system chatter and vibration—equipment should be properly supported to avoid vibrations that could shake loose particulates in the system. Additionally, improperly sized valves—that is, valves that are over-sized for the application, can chatter instead of remaining fully open, which can generate particles.
- Consider source isolation—a good safety measure is to provide means of isolating the oxygen source in a system. Eliminating the oxidizer eliminates one leg of the fire triangle, and can minimize damage if ignition and combustion does occur.
- Use oxygen compatible materials—this will be dealt with in the Material Compatibility section of this chapter, but suffice it to say that material selection is one of the most important aspects of oxygen system design. Many materials can be used safely, some materials can be used safely depending on the particular design parameters, and some materials are entirely unsuitable for use in oxygen system design.

5.10 Design of Liquid Oxygen (LOX) Systems

In addition to design features recommended for GOX system, there are some additional considerations that should be made when designing equipment and systems for cryogenic LOX service.

- The design of liquid oxygen bulk systems storage systems should include the following features:
  - Installation should be above ground, outdoor, or in fire-resistive, non-combustible building.

- The area should be adequately vented to prevent possible build-up of oxygen concentration.
- The installation should not be exposed to power lines or flammable or combustible materials & fluids.
- The installation should be separated from other structures & buildings. NFPA 55 provides standards for determining the Quantity/Distance relationship between bulk oxygen systems and other structures (Fig. 5.8 is reproduced from that standard).
- Prevent spillage into drains—until it vaporizes and warms up, liquid oxygen is denser than ambient air. If it spills into open drains, as it warms it could migrate to uncontrolled areas, causing higher oxygen concentration and risk of fire. A good design feature is to consider use of dikes around bulk oxygen systems that can contain any spills until they vaporize and dissipate.
- Limit access—as a general rule, bulk liquid oxygen systems (and any cryogenic system for that matter), should be isolated so that only trained, authorized personnel can have access to them. Typical means of limiting access include fences and walls with lockable doors or gates.
- Electrical design—as with drains, electrical conduits can provide a path for oxygen to migrate to uncontrolled areas—sealing or inerting conduits will provide protection against this. Additionally, electrical equipment should not be exposed to an oxygen rich atmosphere, as electrical energy and potential sparks can result in fire. NFPA 70 provides guidelines on the design of electrical equipment and systems.
- Provide over-pressure protection—as with all cryogenic fluids, over pressure in a vessel or piping can occur when liquid oxygen is trapped with no means of venting, for example between two isolation valves in a pipe. Vaporizing oxygen will result in dramatically increased pressure. A standard method of protecting against this event is installing safety relief valves on any piping or pressure vessel volume that could contain trapped liquid oxygen.
- Flammability limits for O_2/fuels—oxygen is one leg of the "fire triangle". Combining oxygen with a fuel and ignition source can result in combustion, but it is important to recognize that flammability limits are dependent on the concentration of the fuel and oxidizer. For example, consider Fig. 5.9, which shows flammability limits of both hydrogen and methane in oxygen/nitrogen. Note that while methane will burn with in a pure oxygen atmosphere between about 5 and 40% methane concentration, hydrogen will burn in a pure oxygen atmosphere between about 5 and 95% hydrogen concentration. This means that greater care must be taken in hydrogen systems due to the wider range of combustibility of that fuel in oxygen.

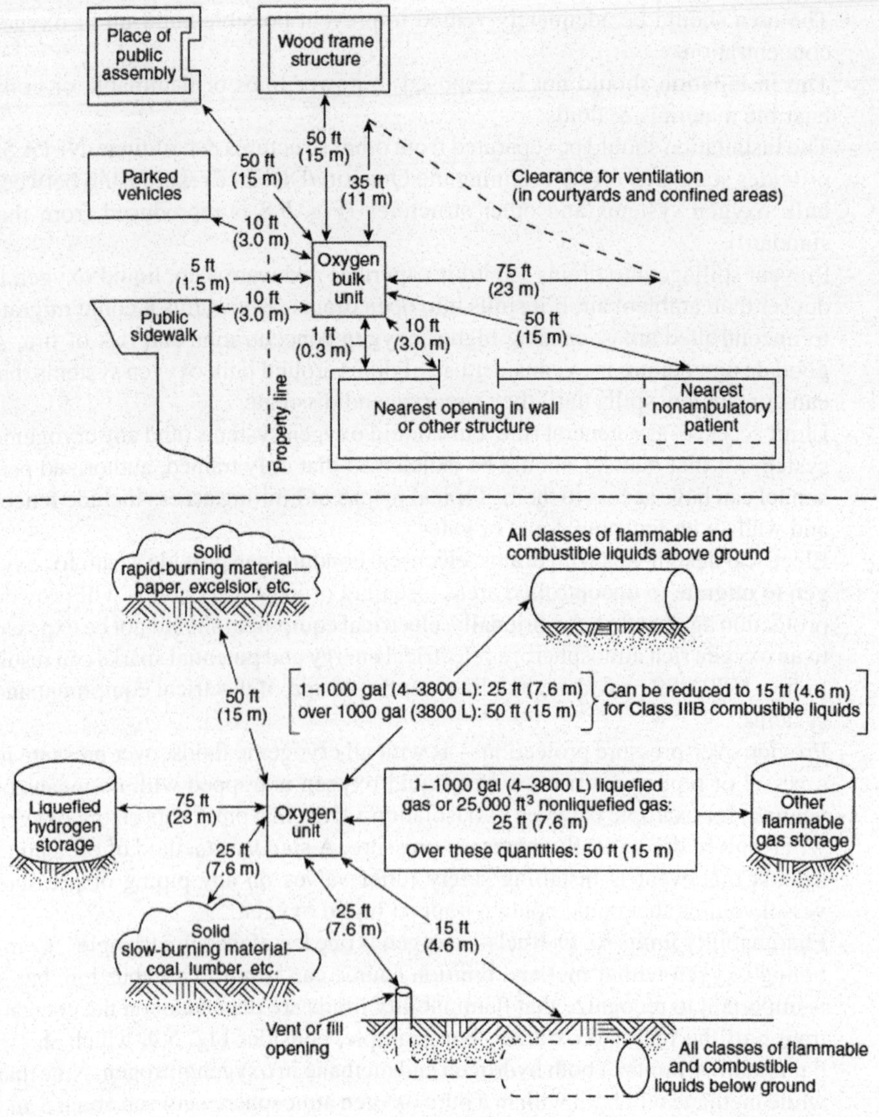

Fig. 5.8 Distance between bulk oxygen systems and exposures (Reproduced with permission from NFPA 55-2016 *Compressed Gases and Cryogenic Fluids Code* © 2015 National Fire Protection Association, Quincy, MA. This reprinted material is not the complete and official position of the NFPA on the referenced subject, which is represented only by the standard in its entirety which may be obtained through the NFPA website at www.nfpa.org)

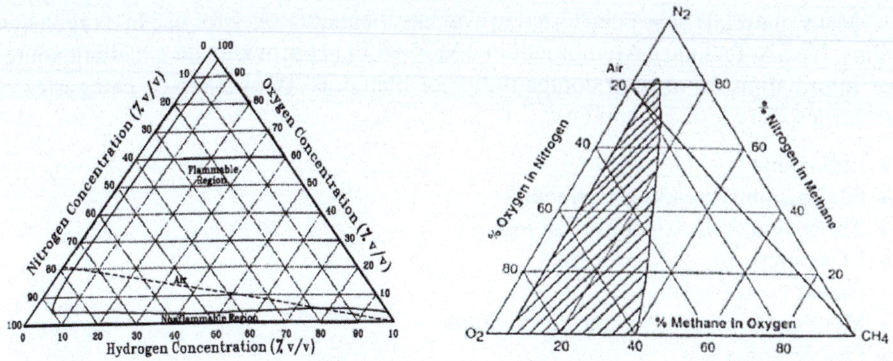

H_2-O_2-N_2 flammable limits [15] CH_4-O_2-N_2 flammable limits [6]

Fig. 5.9 Flammability limits for hydrogen and methane

5.11 Material Compatibility

When designing equipment or systems for use in oxygen, it is obvious that whatever one designs must be fabricated from *something*. However, the dilemma a designer faces is that given the right conditions, pretty much *everything* is flammable. It remains for the designer to choose materials that are most suitable for the design and operating conditions and will minimize the probability of system failure. Some general aspects of material choice to consider are:

- Metal versus non-metal—non-metals are generally more prone to ignition than metals (although metals can also burn). Use of non-metals should therefore be minimized. Where they must be used, in instances such as seals, it is preferable that they not be subjected to wear or movement. As just mentioned, metals can burn and will do so more vigorously in an oxygen-rich environment. The physical configuration of metals also has an influence on their flammability. That is, a copper tube is much less susceptible to ignition than a copper wire mesh.
- Operating limits—choice of materials used in oxygen systems must include consideration of the pressure, temperature, and flow rate that the system is designed for. The higher the pressure and temperature, the less energy is required for a material to ignite than at ambient temperature and pressure. High flow could result in ignition if particles are entrained in the flow and impact part of the system.
- Autoignition is the lowest temperature at which a material will spontaneously ignite without an external ignition source. Generally, the autoignition temperature of typical materials used in oxygen service is relative high (for example—polymers such as TFE© and Kel-F© have autoignition temperatures of over 400 °C). If there is the potential of high temperatures in a system, either due to external sources or heat of compression, a careful evaluation of materials used is necessary.

Many materials have been tested to evaluate their suitability for use in oxygen systems. NASA Technical Memorandum TM X-985 [11] provides an excellent source of information on oxygen compatibility of materials. A number of categories of materials were tested, including:

- Lubricants
- Sealants and threading compounds
- Thermal and electrical insulation
- Elastomers, plastics, adhesives
- Gaskets and packing
- Metals, alloys, solders, surface treatments
- Dye penetrants

Some general rules for material compatibility (not an exhaustive list) are as follows:

Items generally not suitable for use in oxygen systems

- Brittle materials with LOX—brittle transition temperature <−40 C (Carbon steel for example)
- Titanium—although metals are generally suitable, Titanium has some impact sensitivity in oxygen
- Petroleum-derived and silicone-based lubricants, greases & fluids
- Polymer foam insulation, although ceramic foam insulation such as FOAMGLAS© is considered non-combustible
- Mylar—Mylar is a typical aerospace material often used in multi-layer insulation (MLI) systems. Although it is a suitable material for cryogenic applications, as it can burn, it should not be used in oxygen MLI insulation
- Soft solder is generally not an acceptable material for making joints in oxygen systems.

Items generally considered suitable for use in oxygen systems

- Al, Cu & Cu alloys, Ni and Ni alloys (Inconel, Monel), stainless steel
  - A caveat when using aluminum—aluminium is easily ignitable by particle impact. It is generally suitable for containers, but should be avoided for piping, valves & components
  - Iron alloys can be used conditionally when there is no credible ignition source, and operating above −40 °C
- Kel-F©, TFE©, Viton©
- Solder with high silver content is an acceptable material for making joints in oxygen systems
- Many fluorinated & chlorinated fluids & greases.

In all cases, a detailed examination of a system's operating conditions, and a careful evaluation of materials used is necessary for the design of a safe system. Additional information on the use of materials in cryogenics may be found in Chap. 1.

5.12 Construction/Fabrication/Operations

5.12.1 Facility Planning

When planning for oxygen systems and equipment in facilities, it is important to recognize that there are different types—for example public, private, industrial, and medical. Each has it's own unique requirements, and typically different standards/regulations govern how oxygen is used. For example medical oxygen is regulated in the United States by the Food and Drug Administration (FDA) current good manufacturing practices, and by 21 CFR parts 210 & 211. However, a facility that fills high-pressure oxygen cylinders must comply with for example OSHA Standard 1910.253, Compressed Gas Association (CGA)—Safe Handling of Compressed Gases, National Fire Protection Association (NFPA), and the Uniform Fire Code (UFC). A list of applicable standards is included at the end of this chapter.

Installation of oxygen systems should also consider monitors and alarms. As previously pointed out, either low oxygen or high oxygen levels can be dangerous. If there is a risk of either in a confined area, oxygen monitors are essential. When installing monitors, consider where they are located. The closer the monitor is to the source, the earlier the warning. Also, locate monitors where there is good air circulation. If a monitor is located in a dead area with little circulation, it is less likely to provide a timely warning. Additionally, one must consider where do alarm signals go? Locally, or to a remote monitoring location? A local alarm is important to warn people working in an area with potential for high or low oxygen concentration, but it may also be valuable for alarm signals to be sent to remote monitoring areas to facilitate faster response from emergency personnel.

Honoring proper quantity/distance relationships between oxygen system and surroundings is also an essential element of oxygen facility planning. As mentioned in the section titled "Design of Liquid Oxygen (LOX) Systems", NFPA 55 provides guidance on distances between bulk oxygen systems and other equipment and structures. High-pressure oxygen cylinders should be stored separately from high-pressure flammable gasses such as hydrogen or acetylene, and also separated from other toxic or reactive high pressure gas cylinders.

Adequate venting should be provided. Oxygen systems located outside in for example a courtyard should have at least one side open to allow adequate circulation of air. Oxygen systems and equipment located inside should also be adequately ventilated, either with mechanical ventilation, or natural ventilation if it can be verified to be adequate. NFPA 55 specifies 1 $ft^3/min/ft^2$ floor area (0.3048 $m^3/min/m^2$ floor area) of ventilation for rooms where oxygen is stored.

Limit ignition sources. Ignition sources can be static producing equipment, open flames, or electrical equipment where sparking can occur. Grounding is required for equipment with the potential for producing static electricity. It should be obvious that open flames and smoking shall be prohibited near oxygen equipment and systems. If there is a risk of oxygen concentration exceeding allowable limits, electrical equipment in that area should be in accordance with NFPA 70 National Electric Code.

Access to oxygen equipment and systems should be limited to authorized personnel. Equipment should be secured against untrained and unauthorized personnel by use of barricades, fences, and signage.

Good facility design should also minimize the number of joints in the system. Mechanical joints, such as threaded or flanged connections are potential sources for leaks. Good design should maximize the use of welded or soldered connections.

5.12.2 *Manufacturing, Installation and Operations*

When manufacturing, installing and operation oxygen systems, there are some fundamental questions that need to be answered. What are you building, and how will it be used? Is this a single component, or a system? How do you draw the limits around what you are manufacturing? How will it fit into its environment? Who will be using it, sophisticated or unsophisticated users? What level of automation versus human operation will there be? What is its life cycle, and how should it be maintained over its expected lifetime? Answering these questions early in the process will help guide the process and assure satisfactory design and operation.

Materials—previously mentioned was the importance of using oxygen compatible materials. Additionally, strict control is required in managing the use of materials during manufacturing. A system of material certification tracking should be used to assure that the correct materials are being used. Materials should be stored and properly marked to avoid the risk of using the wrong material or component.

Processes used in the manufacture and operation of oxygen systems should be well developed, consistent, traceable, and efficient. Written procedures for material handling, welding, cutting, assembling, cleaning, packaging and storage are necessary to verify that a system is built and operated in a safe and effective manner. Although it is not strictly a requirement for manufacturing, many organizations comply with standards such as ISO 9001, which verify the adherence to a quality system that assures a consistent product. Appropriate administrative controls should be in place to assure that only qualified individuals carry out manufacturing, testing, cleaning and installation of equipment. It should go without saying that pressurized equipment such as piping and pressure vessels that fall under the jurisdiction of codes such as the ASME Boiler and Pressure Vessel code or PED shall only be manufactured by personnel with the appropriate certifications.

Equipment that has been fabricated, tested and cleaned for oxygen service must be appropriately marked and packaged. Details on cleaning and packaging are given in the section of this chapter titled "Cleaning". Storage and distribution shall safeguard equipment against damage and dirt. Packaging for oxygen-cleaned equipment installed in the field should not be opened until immediately prior to installation to protect the equipment from contamination.

Oxygen equipment that has been manufactured and installed at a users site must include instructions for operation and maintenance. Ideally, operation of oxygen systems should include personal, on-site training to assure equipment is operated properly. Written instructions for start up, operation, shut down, and maintenance shall be provided to avoid mishandling of equipment and risk of damage to other equipment or injury to personnel. Instructions should also include emergency procedures in case of an accident. It must be very clear to users what they are and are *not* allowed to do in emergency situations. In the case of personnel injury or fires, it is appropriate for emergency response teams to act. If an oxygen system is installed in a situation where local emergency responders are involved (such as local EMT and fire departments), it is important that they are made aware of the particular risks associated with oxygen systems and are properly trained in response.

During manufacturing and field installation, it is necessary to maintain shop and site cleanliness. Work sites should be clean from flammable materials and dirt that could find their way into oxygen systems. Tools used to fabricate and install oxygen-cleaned equipment must be maintained clean and not be used for work on both oxygen and non-oxygen cleaned equipment.

Personnel working on oxygen systems should always use appropriate PPE. The topic of PPE is covered in detail in the section of this chapter titled "Personnel Safety & Health".

Adherence to the unique requirements of operating oxygen equipment, especially cryogenic oxygen equipment is necessary to assure safe operation. As mentioned previously, only qualified personnel should operate oxygen systems, and they should have appropriate training and certification. Other requirements include:

- Limit pressurization rates—pressurizing equipment too quickly can result in unacceptably high temperatures due to the heat of compression.
- Limit thermal gradients—cooling down cryogenic oxygen systems may induce significant thermal gradients which can cause unacceptably high stresses or binding in equipment due to thermal contraction.
- Analyze LOX pump operation—cryogenic pumps may be damaged if they are operated in a pressure and temperature range where there is the risk of cavitation from boiling liquid. Pumps should be cooled down to operating temperature (usually by flowing cold gas and liquid through the pump) before operation.
- LOX storage vessel contents should be analyzed periodically. As LOX is repeatedly filled and drained from a vessel, levels of contamination could build up to unacceptable levels.
- If LOX must be drained from a cryogenic vessel, it should be vaporized and warmed before venting. It is not good practice to dump LOX, as it will remain denser than ambient air until it warms up, and result in the risk of high oxygen concentration, or possibly LOX spilling into uncontrolled drains where it could migrate to other areas that have no controls or alarms against high oxygen concentration.
- Periodic examinations of oxygen equipment are also important. Equipment should have a regular inspection schedule to record condition and plan for maintenance.

5.13 Cleaning

Cleanliness in oxygen systems is critical. As outlined in previous sections of this chapter, one must use appropriate materials, design systems to minimize risk, and maintain equipment in good working order. However, even if all these criteria are met, there is still a significant risk of failure or accidents if the system or component is not properly cleaned. Equipment must be thoroughly cleaned of debris and hydrocarbons before use. We use the terms "particulates" to quantify debris, and the term "non-volatile residue" of "NVR" to quantify the amount of hydrocarbon. Particulates in a system can be a source of energy if propelled at sufficient velocity and impinging on a surface, and of course any hydrocarbon (NVR) in a system can be a fuel source. There are three things to consider when cleaning equipment for oxygen service. The first is "how clean must it be?" The second is "what method should I use to clean it?" and the third is "how do I verify the cleanliness?"

The first question "how clean must it be?" is not a simple one to answer. "Oxygen clean" is not a single, catchall category. Depending on how and where equipment is used, one can define different levels of oxygen cleanliness. Tables 5.4 and 5.5 [12] show three different standards typically referred to when defining the required level of oxygen cleanliness. It is ultimately up to the engineer to evaluate which standard and which cleanliness level should be used. Several parameters influencing the risk of combustion should be taken into account when determining the level of cleanliness required:

- Temperature—an increase in operating temperature means that a decrease in energy to initiate combustion is required
- Pressure—an increase in operating pressure results in a lower ignition temperature required to initiate combustion
- Concentration—an increase in oxygen concentration increases the likelihood of reaction
- Contamination—increased contamination (NVR) provides a potential source of fuel and particulates can provide a source of ignition in high velocity streams
- Heat of compression—if equipment is subject to high pressures, heat of compression can result in sufficient energy to initiate combustion.

The second question "what method should I use to clean it" depends not only on the final cleanliness requirement, but also the original condition of the equipment. Typical steps for cleaning are:

- Initial degreasing—This is required especially for heavily oil or grease contaminated components. Vapor degreasing is a typical procedure that works well for degreasing heavily contaminated components
- Mechanical cleaning—Equipment with obvious rust, scale and dirt can be mechanically cleaned by wire brushing, scraping, abrasive blasting, or high pressure water. Care must be taken to assure that all debris from mechanical cleaning is removed before further cleaning. Also note that mechanical cleaning is not suitable for

Table 5.4 Maximum allowable particles for various cleaning levels

Cleaning test level	NASA specifications		ASTM G 93 specifications		CGA specifications[b]	
	Particle size range, μm	No. of particles per 0.1 m^2, 1 ft^2	Size range, μm/100 mL	No. of particles allowed per 0.1 m^2, 1 ft^2	Particle size range, μm	No. of particles per 0.1 m^2, 1 ft^2
Typical maximum allowable particles for various cleaning levels						
1000	<500	Unlimited[a]	–	–	500–1000	2.15
	500 thru 750	34	–	–	>1000	0
	>750 thru 1000	5	–	–	Fibers[c]	–
	>1000	0	–	–	–	–
500	<100	Unlimited[a]	x < 100	No limit	–	–
	100 thru 250	1075	100 < x < 175	100	–	–
	>250 thru 500	27	175 < x < 300	20	–	–
	–	–	300 < x < 500	5	–	–
	>500	0	x > 500	0	–	–
	–	–	Fibers		–	–
300	<100	Unlimited[a]	x < 100	No limit	–	–
	100 thru 250	93	100 < x < 175	20	–	–
	>250 thru 300	3	175 < x < 300	5	–	–
	>300	0	x > 300	0	–	–
	–	–	Fibers	25	–	–
200	<50	Unlimited[a]	–	–	–	–
	50 thru 100	154	–	–	–	–
	>100 thru 200	16	–	–	–	–
	>200	0	–	–	–	–
175	–	–	x < 50	20	–	–
	–	–	50 < x < 100	5	–	–
	–	–	100 < x < 175	1	–	–
	–	–	x > 175	0	–	–

(continued)

Table 5.4 (continued)

Cleaning test level	NASA specifications		ASTM G 93 specifications		CGA specifications[b]	
	Particle size range, μm	No. of particles per 0.1 m^2, 1 ft^2	Size range, μm/100 mL	No. of particles allowed per 0.1 m^2, 1 ft^2	Particle size range, μm	No. of particles per 0.1 m^2, 1 ft^2
	–	–	Fibers	5	–	–
100	<25	Unlimited[a]	–	–	–	–
50	25 thru 50	68	–	–	–	–
	>50 thru 100	11	–	–	–	–
	>100	0	–	–	–	–
	<10	Unlimited[a]	–	–	–	–
	15 lhru 25	17	–	–	–	–
	>25 thru 50	8	–	–	–	–
	>50	0	–	–	–	–

[a]"Unlimited" means particulate in this size range is not counted; however, if the accumulation of this silt is sufficient to interfere with the analysis, the sample shall be rejected
[b]Cleaning Equipment for Oxygen Service, CGA G-4.1, Compressed Gas Association, Arlington VA, 1996
[c]Isolated fibers of lint shall be no longer than 1000 μm, and there shall be no accumulation of lint fibers

Table 5.5 Nonvolatile residue level specifications

Level	NASA		ASTM G 93		CGA specifications[b]	
	Specifications		Specifications		Acceptable contamination	
	Maximum quantity NVR		NVR remaining		Level	
	mg/m^2	lb/ft^2	mg/m^2	lb/ft^2	mg/m^2	lb/ft^2
Typical nonvolatile residue (NVR) level specifications						
A	10	2.05×10^{-6}[a]	<11	$<2.25 \times 10^{-6}$	–	–
B	20	4.10×10^{-6}	<13	$<6.76 \times 10^{-6}$	–	–
C	30	6.14×10^{-6}	<66	$<1.35 \times 10^{-5}$	–	–
D	40	8.19×10^{-6}	<220	$<4.51 \times 10^{-5}$	–	–
E	–	–	<550	$<1.13 \times 10^{-4}$	500	$<1.02 \times 10^{-4}$
F	–	–		Specified by user or supplier	–	–

[a]NVR level commonly specified for NASA oxygen systems
[b]Cleaning Equipent for Oxygen Service, CGA G-4.1, Compressed Gas Association, Arlington VA, 1996

non-metallic materials, and blasting is not recommended for aluminium alloys. If using a wire brush, make sure the bristles are not made of carbon steel.

- Disassembly—Equipment should be disassembled as much as practical to ensure each area of the component is thoroughly cleaned. Additionally, parts should be grouped according to cleaning method used. For example, some parts may be immersed in cleaning solution, where some others may be flushed or sprayed.
- Pre-cleaning—Equipment can be washed with commercially available cleaning solutions. Cleaning solutions fall into the categories of solvents, detergents, alkaline (caustic) solutions, or acids. Solvents are becoming less common due to health and environmental concerns, and are more recently replaced by other solutions. Acid solutions are effective for removing rust and oxides from surfaces, but should be used with caution, as they are not always the best choice for degreasing equipment. Additionally, they may cause stress corrosion in certain metals. Caustic solutions are generally effective for degreasing equipment, and are typically used at elevated temperature to improve effectiveness. Finally, detergents are entirely suitable for oxygen cleaning, and are just as effective as other methods for degreasing equipment. For all cleaning solutions, it is critical that after cleaning, equipment must be thoroughly rinsed with clean (de-ionized, distilled or filtered) water and allowed to dry. Flushing or blowing equipment with clean (oil free) dry air, or an inert gas such as nitrogen will help the drying process. Alternatively, drying can also be done using vacuum or elevated temperature.
- Precision cleaning—if equipment requires a more stringent cleaning specification, precision cleaning may be done using a vapour degreaser, or ultrasonic cleaning. As with cleaning solutions, equipment must be thoroughly rinsed and dried.

The third question "how do I verify the cleanliness" is just as important as the cleaning protocol. There is not a single cut-and-dried answer, but a number of methods that can be used to verify cleanliness. Verification can be done either by inspection, or by process (or a combination of the two). The acceptance criteria that the designer chooses informs what type of inspection should be performed. Typical inspection methods and their threshold detection limits are shown in Table 5.6 [13]:

A visual inspection with bright light can be used to see if there are any rust, metal chips, weld slag, grease, oil, paint, or moisture. This method is very rudimentary, and should be used as a first look before making further inspections.

Table 5.6 Detection methods and threshold detection limits

Test method	Informative threshold of detection (mg/m^2)
Bright white light	500–1700
UV (black light)	40–1500
Wipe test	30–600
Water break test	30–60
Solvent extraction	<10

Inspection with UV (black light) does have limited value in detecting the presence of oils and greases. However, caution should be used, as some hydrocarbons do not fluoresce under UV light and may go undetected.

A wipe test is also a simple way to check cleanliness level. A white filter paper or lint free cloth is wiped over the surface being tested and examined visually. Any particulates or stains from oils can be seen more easily than by the direct visual inspection method.

The water break test is a good way to check for the presence of oils and greases. Clean distilled water sprayed on a horizontal surface will tend to bead up if there is oil or grease present on the surface.

The most precise inspection method is solvent extraction. In this method, a known surface area of the item to be inspected is flushed with a quantity of solvent such as IPA. The solvent can then be filtered through a filter paper, which was previously weighed. The solvent is evaporated, and filter re-weighed to determine the amount of NVR remaining. Additionally, the filter can be examined under magnification to count the number of particulates.

Another issue to consider is how much of the equipment should be inspected. If it a relatively small component, the entire surface can be inspected. However, for large systems, take for example a large vessel or piping system, it is impractical to inspect the entire surface. In this case, a representative sample area can be inspected.

Procedure—use a validated process! Inspection is a required step, but can only validate that the equipment has been adequately cleaned. A qualified procedure is the best guarantee of oxygen cleanliness.

Once equipment has been appropriately cleaned for oxygen service and inspected, it is Important to maintain the cleanliness level up to and including the time that it is installed and used. *Packaging* is an important step in the oxygen cleaning process. After equipment has been cleaned and inspected, it should be sealed from contaminants. Small parts are typically sealed in plastic bags, where dust caps, blind flanges and waterproof tape can be used to seal openings in larger equipment.

Ideally, smaller parts should be double bagged, where the inner bag shall be as clean as the part being packaged. An outer bag is then used as additional protection from dust and moisture. Larger equipment such as pumps and piping should be purged with clean dry air or nitrogen to remove atmospheric air, which could contain moisture. If the equipment can be sealed against internal pressure with plugs or blind flanges, a slight over-pressure can be used to assure that atmospheric air does not ingress. In this case, it is important to include a warning that the equipment is under pressure and care must be taken before removing any seals.

In all cases, the equipment should be *labeled* to indicate at a minimum that the item has been cleaned for oxygen service, to what specification it has been cleaned, and the date it was cleaned and sealed. Equipment packaging should not be opened until equipment is ready to install.

5.14 Oxygen Hazards Examples (Informative)

5.14.1 Example 1—Suitability of G10 for Instrumentation Support in a LOX Cryostat

Introduction

G-10 micarta is a common thermosetting industrial laminate consisting of a continuous filament glass cloth material with an epoxy resin binder. This product has the characteristics of high strength, excellent electrical properties and chemical resistance. This material has been used in cryogenic systems as a structural and electrically insulating platform for mounting instrumentation inside cryogenic dewars.

In this example, it is proposed to use G-10 as a structural support for an instrument inside in a research cryostat in oxygen service. Although micarta is not an ideal material for use in oxygen [1], it needs to be determined if it acceptable for use in this specific application.

The following information provides justification and technical background supporting use of G-10 micarta for the subject application.

Given

- G10 support mass—0.2 kg
- Cryostat internal depth—2 m
- Initial pressure—1 bar(a)
- Final pressure—6 bar(a)
- Operating temperature—minimum 90 K, maximum 293 K
- Instrument power—30 mA at 10 V

G10 Material Properties

G10/FR4 is the flame retardant version of **G10**. Properties are shown in Table 5.7 [2]:

Oxygen Compatibility

Determining compatibility of materials with oxygen is not a cut-and-dried "go" or "no go" affair. Materials must be assessed for specific applications considering the "Fire Triangle". First of all, one must consider the availability of fuel (in this case—does the proposed G10 material make a good fuel?). Secondly, an oxidizer must be present. As the application is an oxygen cryostat, it is obvious that the support will be in a 100% oxygen atmosphere. Thirdly, one must consider the potential ignition sources. All three (fuel, oxidizer & ignition) must be present for a hazard to exist. The contribution of these three factors is evaluated, and the overall risk determined as follows:

Potential Failures

1. Electrical short causing arc & ignition of G10.
2. Failure of attachment causing impact of G10 on cryostat wall.

Table 5.7 G10/FR material properties

Density	0.065 lbs/in^3 (1.8 g/cm^3)
Maximum operating temperature	–
Electrical	266 °F (130 °C)
Mechanical	284 °F (140 °C)
Dielectric strength	800 kV/in (31.5 kV/mm)
Dielectric constant (@ 10^6 cycles per second)	5.0
Volume resistivity	6×10^6 M Ω cm
Arc resistance	100 s
Surface resistivity	1×10^6 M Ω
Parallel dielectric	60 kV
Auto-ignition temperature [3]	258 °C
Heat of combustion [4]	10,440 J/g (4486 BTU/lb)
Highest passing mechanical impact	–
Energy in LOX at ambient pressure	16.3 J (12 ft lb)
Pressure in GOX	0.5 MPa (72.5 psia)

3. Adiabatic compression causing increase of G10 temperature above auto-ignition temp.
4. Failure of other components causing impact on G10.

Analysis

1. The instrument is powered by a current source that provides 30 mA maximum current at a compliance voltage of 10 VDC. If the instrument fails due to a short circuit and the current arcs across the power leads, the amount of power delivered would be 0.03A × 10 V = 0.3 W, or 0.3 J s. Per Table 5.7, G10 can withstand 16.3 J of mechanical energy without failure. So, every second, the maximum Joules from a short would be 0.3 J, which is <2% of the amount of energy that G10 can tolerate.
2. Another potential failure would be due to the G10 becoming unattached from its support and falling to the bottom of the tank. Given the support mass of 0.2 kg, if we assume all the mass is somehow concentrated in a compact volume, and falls the entire depth of the cryostat (2 m), it's final velocity at impact would be 6.3 m/s, and it's kinetic energy would be (m * V^2)/(2*g_c) = 0.4 kg m = 3.9 J. Again—this is much less (approximately 24%) of the amount of energy that G10 can tolerate without failure.
3. Adiabatic compression causing increase of G10 temperature above auto-ignition temperature can be calculated using the following equation:

$$T_f = T_i * \left(\frac{P_f}{P_i}\right)^{\frac{k-1}{k}}$$

where

T_f final oxygen temperature
T_i initial oxygen temperature
P_f final tank pressure
P_i initial tank pressure
k specific heat ratio = 1.4 for oxygen

Assuming the worse case,

T_i ambient temperature = 293 K
P_i 1 bar(a)
P_f 6 bar(a)

And therefore

T_f 489 K

Per Table 5.7, the auto-ignition temperature of G10 is 531 K, which is greater than T_f, so the G10 would not auto-ignite. Also, one can consider that adiabatic compression assumes that no heat escapes from the compressed oxygen. In reality, the cryostat wall provides a significant heat sink, which makes the compression more isothermal, and in reality the final temperature of the oxygen would be less than 489 K.

4. Failure of other components impacting the G10 could result in high enough energy to cause the G10 to combust. Presumably, this scenario could happen if there were components in the system that might come loose during a rapid decompression of the cryostat, or a major failure like the cryostat coming loose from its supports. Good design should minimize this risk. However, in the event that a component could impact the rake causing complete combustion, the impact on the tank must be evaluated. Assuming that the G10 does start on fire, the flammability of the cryostat material (stainless steel) is assessed by determining if the burning G10 would initiate sustained combustion of the stainless steel, or if it would simply burn out without igniting the stainless steel.
This scenario most closely resembles the Promoted Combustion test (ASTM G 124). The test determines the ability of a metallic rod to propagate flame upward when ignited at the bottom by an ignition source.
Mass of the G10 support is given as 0.2 kg. G10 heat of combustion is 10,440 J/g. Should the entire G10 support be consumed, it would generate about 2088 kJ. However, this is not particularly significant, as the quantity of promoter (burning material) does not affect the metal flammability. For any metallic material, the flammability increases with increasing pressure. The threshold minimum pressure required to support self-sustained combustion in a 300 series stainless steel rod per ASTM G 124 is ~68 bar(a). The maximum working pressure of the cryostat is given as 6 bar(a). We therefore conclude that even if the G10 did completely combust, conditions are not sufficient for sustained combustion of the stainless steel support. The G10 would simply burn itself out. It should also

be noted that the ASTM test more severely evaluates metallic materials because the aluminum promoter used in the test is a far more intense ignition source than typical ignition sources in real systems (such as G10 would be).

Conclusion

G10 micarta is an acceptable material to fabricate an instrument support for the subject application. Ignition hazards from arcing, impact and adiabatic compression are small enough or can be controlled with system alarms to prevent this material from causing any significant failure of the cryostat.

References

1. NASA Technical Memorandum TM X-64711 (N74-76467)—"Compatibility of Materials with Liquid Oxygen Volume 1", C.F. Key, 10/1/72.
2. G10 properties MIL SPEC 24768/27.
3. ASTM G63 Table X1.2.
4. Flammability and Sensitivity of Materials in Oxygen-Enriched Atmospheres, Vol. 8, ASTM STP 1319, p. 425.
5. ASTM Manual MNL36—"Safe Use of Oxygen and Oxygen Systems", Beeson et al., 2000.

5.14.2 *Example 2—Oxygen Enrichment Due to Un-insulated Cryogenic Piping*

Introduction

An ambient heat exchanger is located in a room that contains cryogenic equipment (see Fig. 5.10). The heat exchanger warms up cryogenic helium from approximately 4–300 K. There is a risk of oxygen enrichment due to the preferential condensation of oxygen from the atmosphere, as the vapor-to-liquid transition temperature for oxygen is slightly warmer than nitrogen. This example provides a first-order analysis of the potential risk and recommended mitigation strategies.

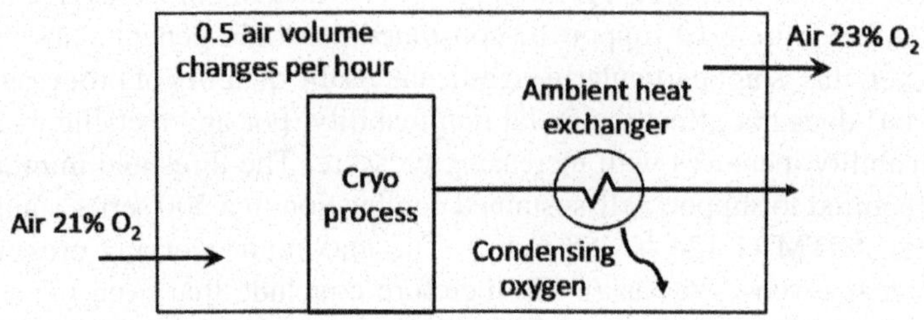

Fig. 5.10 Example problem oxygen enrichment

Assumptions

The Compressed Gas Association (CGA) defines oxygen-enriched mixtures or atmospheres as any mixture or atmosphere containing greater than 23% oxygen, since above this concentration, the reactivity of oxygen significantly increases the risk of ignition and fire. Therefore, we consider maximum allowable concentration in the cold box room to be 23% O_2 by volume.

Room size—approximately 65 × 12.5 × 13 m
Cold box room volume ~10,600 m^3
Air temperature = 300 K
Atmospheric pressure = 1 bar
Room air changes per hour = 0.5/h
Ambient heat exchanger capacity 30 kW

Analysis

As a first order analysis, we consider how much oxygen would have to be condensed to increase the oxygen content in the room to 23%.

Given air composition is normally 21% O_2, 78% N_2, ~1% Ar.

Partial pressure O_2 = 0.21 bar
$\rho_{(O2)}$ at 300 K, 0.21 bar = 0.2694 kg/m^3
Mass of O_2 in room at 21% concentration
= 10,600 m^3 × 0.2694 kg/m^3 = 2856 kg

To increase the O_2 concentration to 23% we calculate:

$\rho(O_2)$ at 300 K, 0.23 bar = 0.2951 kg/m^3
Mass of O_2 in room at 23% concentration
= 10,600 m^3 × 0.2951 kg/m^3 = 3128 kg

The increase in O_2 mass required to increase the O_2 concentration to 23% is therefore:

= 3128–2856 = 272 kg

We see then, that for every air change in the cold box room we would have to condense 272 kg O_2 to increase its concentration to 23%.

The O_2 condensation rate is then:

0.5 air change/h × 272 kg O_2/air change = **136 kg/h.**

Therefore, we would have to condense 136 kg/h of O_2 to increase the oxygen concentration in the entire room to 23%.

Next we determine how much energy must be removed from the air to reduce its temperature from 300 K to the boiling point of oxygen and then condense the oxygen.

O_2 b.p. at 0.21 bar = 77.4 K
N_2 b.p. at 0.78 bar = 75.2 K

Ar b.p. at 0.01 bar = 83 K

So, if we reduce the air temperature at the vaporizer to 77 K, we will condense oxygen and argon, and just have cold nitrogen vapor.

Enthalpy change and phase change O_2 300 K at 77 K

$\Delta h = (273–69.5) = 203.5$ kJ/kg
$h_{fg} = 224.5$ kJ/kg
Total = 428 kJ/kg

Enthalpy change and phase change Ar 300 K at 77 K

$\Delta h = (158–43.6) = 114.4$ kJ/kg
$h_{fg} = 162$ kJ/kg
Total = 276.4 kJ/kg

Enthalpy change N_2 300 K at 77 K (no phase change)

$\Delta h = (314–75.5) = 238.5$ kJ/kg
Total = 238.5 kJ/kg

For every kg of air, we have (based on the density of Ar, N_2, O_2 at their respective partial pressures):

0.015 kg Ar
0.23 kg O_2
0.755 kg N_2

To reduce 1 kg of air from 300 K to 77 K, and condense O_2 and Ar, takes:

0.015 kg × 276.4 kJ/kg = 4.1 kJ (Argon)
0.23 kg × 428 kJ/kg = 98.4 kJ (Oxygen)
0.755 kg × 238.5 kJ/kg = 180.1 kJ (Nitrogen)

Total heat to be removed = 282.6 kJ per kg of air.

This means that for each kW of refrigeration, 98.4/282.6 = 0.348 kW goes into cooling and condensing the oxygen.

The ambient heat exchanger cooling capacity = 30 kW. Therefore, the most oxygen that could be cooled and condensed (not counting heat exchanger efficiency) is:

30 kW = (30 kJ/s) × (0.348 kW/kW) × (1 kg/428 kJ) × (3600 s/h) = 88 kg/h

Therefore the maximum condensation rate of oxygen of 88 kg/h is less than the required condensation rate of 136 kg/h to raise the concentration of oxygen in the room to 23%.

Summary

We see from the preceding analysis, that to a first order approximation, the maximum oxygen condensation rate would be only approximately 2/3 of that required to raise the oxygen level in the room to 23%, which is considered the level where there is increased risk of ignition and fire. *Note that this analysis assumes well-mixed air in the entire cold box room.* Obviously, the local concentration of oxygen could be increased if this assumption is invalid.

Recommendations

The following recommendations should be considered:

- Keep any ignition sources away from the ambient heater that could generate electrical sparks, grinding or naked flames (hot work, cigarettes, etc.). In other words any activity and equipment in the vicinity should be limited. This also applies to combustible materials (e.g. Oil).
- Operators working near the ambient heater should be given clear safety instructions before starting their activity.
- Local fire extinguishing equipment should be located near the oxygen-enriched areas.
- Warning signs (oxidizing agents) should be affixed close to the areas concerned by oxygen-enrichment.
- A local O_2 monitor could be installed near the ambient heater.
- Provide a drip pan under HX8 to catch and contain any O_2 condensation.
- Consider alarms upon loss of room air ventilation.

5.15 Best Practices

1. Consideration must be made to address separately the unique hazards of both GOX and LOX.
2. Make appropriate separation between oxygen systems and sources of fuel and ignition—remember the “fire triangle”
3. Use approved containers for storage and transportation
4. Oxygen systems should have a defined maintenance plan to assure continued safe use
5. Training in the use of oxygen systems is important, as there are many unsophisticated users of oxygen systems
6. There are many well developed standards for oxygen system design—use them
7. Pay attention to material compatibility issues
8. As with all cryogenic system, always use appropriate PPE when working with oxygen systems
9. Proper cleaning of components and systems in oxygen service is critical.

5.16 Standards and References

The following is a list of useful references related to various aspects of the design, fabrication, installation, and operation of oxygen equipment and systems. Note that in some cases, the references listed may be obsolete, but are still shown, as they may not have been replaced with more current documents. Always check to find the most recent edition of a reference.

5.16.1 General

- Safe Use of Oxygen and Oxygen Systems—Handbook for Design, Operation, and Maintenance (2nd Edition), Beeson et al., ASTM International, ISBN 978-0-8031-4470-5
- ASTM G-88-13 Standard Guide for Designing Systems for Oxygen Service
- CGA G-4 Oxygen (general)
- CGA G-4.3 Commodity Specification for Oxygen
- CGA G-4.4 Oxygen Pipeline and Piping Systems
- CGA has many additional documents on oxygen at:

  https://www.cganet.com/

- NFPA 50 Standard for Bulk Oxygen Systems at Consumer Sites
- NFPA 53 Recommended Practice on Materials, Equipment, and Systems Used in Oxygen-Enriched Atmospheres
- ASTM G63 Standard Guide for Evaluating Nonmetallic Materials for Oxygen Service
- ASTM G88 Standard Guide for Designing Systems for Oxygen Service
- ASTM G94 Standard Guide for Evaluating Metals for Oxygen Service
- EIGA Doc. 127/13 Bulk Liquid Oxygen, Nitrogen and Argon Storage Systems at Production Sites
- EIGA (European Industrial Gases Association) has many additional documents on oxygen at:

  https://www.eiga.eu

- ISO 14624-4 Space systems—Safety and compatibility of materials—Part 4: Determination of upward flammability of materials in pressurized gaseous oxygen or oxygen-enriched environments
- SAE AIR822C Oxygen Systems for General Aviation
- SAE has many additional documents on oxygen at:

  http://www.sae.org/

- NASA TM-X 985 Compatibility of Materials with Liquid Oxygen

- OSHA.gov—accident database for compressed gas
- 29 CFR 1910.104—Code of Federal Regulations Hazardous Materials—Oxygen (OSHA website).

5.16.2 Testing

- ASTM G72 Standard Test Method for Autogenous Ignition Temperature of Liquids and Solids in a High-Pressure Oxygen-Enriched Environment
- ASTM G86 Standard Test Method for Determining Ignition Sensitivity of Materials to Mechanical Impact in Ambient Liquid Oxygen and Pressurized Liquid and Gaseous Oxygen Environments.

5.16.3 Safety

- GRC Safety Manual BMS Document # GLM-QS-1700.1, Chapter 5—Oxygen Document Number: GLM-QS-1700.1.5
- ISO 22538-1,2,3 Space systems-Oxygen safety
- NASA report NSS 1740.16 January 1996 Safety Standard for Oxygen and Oxygen Systems (obsolete)
- EIGA 44/09/E Hazards of Inert Gases and Oxygen Depletion
- EIGA Doc. 04/09 Fire hazards of oxygen and oxygen enriched atmospheres
- U.S. Fire Administration/Technical Report Series Special Report: Fires Involving Medical Oxygen Equipment *USFA-TR-107/March 1999.*

5.16.4 Cleaning

- CGA G-4.1 Cleaning Equipment for Oxygen Service
- ASTM G93 Standard Practice for Cleaning Methods and Cleanliness Levels for Material and Equipment Used in Oxygen-Enriched Environments
- NASA MSFC Spec 164: Cleanliness of Components for Use in Oxygen, Fuel, and Pneumatic Systems (contains acceptable methods of cleaning pipe, tubing, and flex hose)
- NASA KSC-C-123: Specifications for Surface Cleanliness of Fluid Systems.

References

1. http://www.cdc.gov/mmwr/preview/mmwrhtml/mm5731a3.htm
2. P.M. Ordin, NASA TM X-67953 Mishaps with Oxygen in NASA Operations (1971)
3. Y.A. Freiman, H.J. Jodl, *Solid Oxygen*, vol. 401, no. 1–4 (Physics Reports, November 2004), pp. 1–229
4. NIST thermophysical properties, http://webbook.nist.gov/chemistry/fluid/
5. NIST Reference Fluid Thermodynamic and Transport Properties Database (REFPROP): Version 9.1, https://www.nist.gov/srd/refprop
6. T. Flynn, *Cryogenic Engineering*, 2nd edn. (CRC Press, 2004). ISBN 9780824753672
7. Safe location of oxygen and inert gas vents, EIGA document 154/16
8. NFPA 53 "Recommended Practice on Materials, Equipment, and Systems used in Oxygen Enriched Atmospheres"
9. J.M. Clark, The toxicity of oxygen. Am. Rev. Respir. Dis. **110**(6 Pt 2), 40–50 (December 1974)
10. Oxygen pipeline and piping systems, EIGA Document 13/12/E
11. NASA TM-985 Compatibility of Materials with Liquid Oxygen (1964)
12. H. Beeson, S. Sarah, W. Stewart, *Safe Use of Oxygen and Oxygen Systems: Handbook for Design, Operation, and Maintenance,* 2nd edn. (ASTM International, 2007). ISBN 978-0-8031-4470-5, Electronic-ISBN 978-1-61583-677-2
13. AIGA document 12/04 Cleaning of Equipment for Oxygen Service
14. https://www.osha.gov/dte/grant_materials/fy07/sh-16634-07/mod_6_confined_space2.pptx
15. "Guide to Safety of Hydrogen and Hydrogen Systems", AIAA G-095A 2017
16. NASA/TM–2004-213400 Low Earth Orbital Atomic Oxygen Interactions with Spacecraft Materials B. Banks et al, 2004

Chapter 6
Hydrogen Safety

Stephen S. Woods

Abstract Hydrogen Safety is a comprehensive summary of basic safety acumen acquired over the last 60 years from industry and aerospace practice. It is intended to inform technicians, engineers, scientists, and managers who find themselves with the prospect of working in a hydrogen environment in which they have had little experience. It is also set up to provide more experienced hydrogen workers with reminders of how to continue to be safe. Topics introduced include an explanation of the challenges of working with Hydrogen, the basics of how hydrogen hazards are addressed, basic handling, and a compilation of basic component, system and facility considerations. A "Safety Checklist" is provided to be used as suggested, a review of important issues, or to stimulate thinking about how to organize thoughts for setting up a hydrogen activity. Today hydrogen interests are expanding enormously with an explosion of applications. Unfortunately, there have been a spate of serious and deadly accidents involving work with hydrogen across the globe, especially by researchers and developers, that could have been avoided if basic hazards had been better understood. It hoped the information provided in Hydrogen Safety reaches those in need.

6.1 Example Accident

In 1980 at Los Alamos National Laboratory in New Mexico an explosion shook a gas supply facility and hurled shrapnel into property managed by the City of Los Alamos. Two Los Alamos employees suffered extensive burns that would result in 4 months of lost time. City officials demanded an explanation and assurances of safety. The Press dogged laboratory management. The ensuing enquiry found laboratory management grievously at fault for being unaware of the improper hydrogen practice at the gas depot, and for their failure to provide adequate hydrogen training to line management and technicians.

What led to this Accident?

T. J. Peterson and J. G. Weisend II, *Cryogenic Safety*, International Cryogenics Monograph Series, https://doi.org/10.1007/978-3-030-16508-6_6

Fig. 6.1 Close up view of tube bank, post-accident [4]

The facility operation refilled gas cylinders for distribution to the various Los Alamos facilities. Technicians operated a bank of valves supplying different pressurized gases through a common manifold. Their inventory included hydrogen and oxygen. The operators had an *improvement* idea for operations and consulted their line manager. Removal of an isolation valve on the manifold would result in a time savings as one less valve to open. Without hydrogen training the idea seemed sound. The alteration was performed without review by knowledgeable staff or management overview and approval. Immediately upon introducing oxygen into the manifold that retained unvented hydrogen, a mini-explosion resulted that in turn blew out the valve to the tube trailer. An ignition source wasn't necessary. Hydrogen and oxygen in the right proportions can ignite spontaneously. The oxygen supply with its higher pressure surged into the hydrogen within the trailer. This single point failure led to catastrophic results (Fig. 6.1).

Forensic analysis determined that as the injected oxygen mixed with the hydrogen the higher oxygen pressure adiabatically compressed the mixture, raising its temperature until it exceeded the autoignition temperature, and in an instant, accelerated flames formed and transitioned to detonation at a pressure determined to be 550 psi. The tubes, now subject to detonation, ruptured, and with the consequences that large pieces of shattered steel became shrapnel and a significant cloud of hydrogen was released to form a large fire ball over the parking area.

Two employees who happened to be in the parking lot, were caught by surprise and had no chance to evade injury. Miraculously the shrapnel missed them, but they suffered flash burns over a large portion of their bodies. Hydrogen flash fire combusts very quickly, releasing all its energy in a second or two. Any protective action by the victims would have occurred after exposure to the radiant energy had already happened. One massive piece of shrapnel travelled a quarter mile passing the laboratory perimeter into city property, an event that really caught the public's attention. Amazingly, and fortuitously the resting spot of the shrapnel was in the city dump! (Figs. 6.2 and 6.3)

Fig. 6.2 View of the scene of the accident from the parking lot [4]

Fig. 6.3 Shrapnel thrown into the City of Los Alamos [4]

This incident underscores several key points regarding hydrogen safety:

- Changes to any system require careful review. Even minor alterations to a hydrogen system or operation can lead to drastic changes in outcome.
- Combustion in confined volumes, in this instance the tubes of the trailer, can lead to high overpressures capable of shattering heavy-duty steel.
- Management responsibility extends to all aspects of hydrogen operations, and certainly those involving hazardous operations. Ultimately, in this accident Los Alamos management was held accountable.
- Appropriate training of staff is critical.

6.2 Acronyms, Terminology and Definitions

The following abbreviations and terms appear in the text.

Abbreviation	Term	Explanation
CJ	Chapman-Jouguet	Denotes a detonation state defined by theoretical thermodynamic parameters
NBP	Normal boiling point	Boiling point at one atmosphere of pressure
NTP	Normal temperature and pressure	20 °C (293.15 K, 68 °F) and 1 atm (101.325 kN/m^2, 101.325 kPa, 14.7 psia)

6.3 Introduction

These notes summarize issues where care is required in performing work with hydrogen systems. The intent is to inform those new to hydrogen where to focus their attentions in a hydrogen work environment, and for experienced engineer, scientist or technician to provide a review and checklist to avoid missing an important point. For those who support hydrogen work, either in management, procurement, or other support activities, it can serve to inform. The content may be applied to a variety of hydrogen activities including facility planning, laboratory work, operations, test activities, production, manufacture, use of hydrogen appliances, design work and training, to name a few.

Work in your environment will require training and operational protocols specific to the activity involved. The information provided in this text will provide the background to support that work. "Safety Notes" in brackets are placed wherever considerations warrant extra attention. This text is not a comprehensive source, therefore in addition to the basic physical data and practice information provided other sources of information will be referenced.

In this modern era the applications involving use of hydrogen are multiplying rapidly and with new technologies. This work as it exists today is part science, part engineering and part art, and there is no wrong or right way of application. However, hydrogen is a very energetic fluid, and safety must be considered as an integral part of design and application. One of the early pioneers in our modern era, D.B. Chelton wrote [1], "As in any safety program, an explicit set of adequate rules cannot be given; instead, the general criteria must be studied carefully to ensure that every potential hazard has been considered", and this is still excellent guidance for today.

6.4 Primary Safety Issues with the Use of Hydrogen

The characteristic behaviors noted in over a century of hydrogen use make clear that safety considerations must be applied in measure equal to any other consideration whether in planning, design, fabrication, installation, operations, disposal, maintenance, and even in shutdown and termination phases of projects or facilities. Hydrogen, while ubiquitous in the universe at large, does not exist in a free state in our natural environment. Hydrogen employed in the myriad applications found today as well as new applications under contemplation must be manufactured, transported to points of use, stored under conditions that meet the requirements of intended applications, and deployed in systems with some care. It is a fair observation that with the accelerating pace of development of new hydrogen applications, hydrogen practice is not a staid static field and may be considered as much art as engineering. There is no one correct way to proceed. It is an artificial material with no *natural* "safe" state, so we the planners, designers, managers, and operators of systems are **all** responsible for using hydrogen safely.

The primary issues with hydrogen use may be summed up with four areas of concern, and in order of importance, are:

1. **Fire and Explosion**. Hydrogen is flammable over a wide range of mixtures, readily ignited by a variety of ignition sources which in certain circumstances have energies so small as to be considered spontaneous, and in the presence of typical confinements such as lines, ducts, or any enclosed volume, possesses a propensity to generate damaging overpressures.
2. **Pressure Systems**. Most hydrogen systems operate at elevated pressures, and therefore have all the hazards that attend pressure systems. See also Chap. 3.
3. **Materials of Construction**. Proper system material selection is required for the conditions of use, especially in systems that operate at low temperatures. Materials must possess appropriate ductility, and design must account for material expansion and contraction over the operating range. Electronic properties in sensors must be suitable for the temperatures of use. In addition, certain metals

are subject to hydrogen attack that can accelerate stress crack growth resulting in loss of strength, and in the worst of conditions cause dramatic material failure. See also Chap. 1.

4. **Protection of Personnel and Surroundings**. Basic practice dictates that personnel and critical equipment be protected from any possibility of exposure to dangerous combustion and overpressures, usually by isolating the hazard. Operations are permitted around equipment with certified pressure ratings in conjunction with the appropriate personnel protective equipment (PPE) provided to mitigate minor chance exposures. Exposures include flame, ultracold surfaces, eye hazards, possible acoustic hazard, dangers associated with pressure equipment, and asphyxiation, especially where equipment or facilities employ purge gases such as helium or nitrogen.

An additional comment is that safety considerations for hydrogen use involve not only the system elements wetted by hydrogen, but they must include adjacent components and hardware which might be exposed should a release enter their surroundings and cause interaction.

The presentation that follows will address these primary issues. The text identifies issues and their hazards, examines their genesis from physical properties or system characteristics, then notes approaches to addressing hazards and safety guidelines. A focus on the profound influence of confinement is used to hopefully aid the reader's intuition on how to break down hydrogen hazard complexity.

Topics discussed:

- The Challenge to Working with Hydrogen
- Addressing Hydrogen Hazards
- System Considerations
- Facilities
- Safety Checklist

In each of these topics you'll see the same issues resurface as they reflect on different areas of interest whether it be design, basic handling, systems, operations or facilities.

6.5 The Challenge to Working with Hydrogen

Hydrogen is a manufactured product not found free in at the earth's surface as it is chemically bound in organic and inorganic compounds. As a gas at one atmosphere it has low density, so to make it a useful commodity it must be compressed to high pressures, or liquified at cryogenic temperatures, or both, and therefore, its handling has all the attendant concerns of pressurized, and/or cryogenic systems. By itself, hydrogen is not reactive, but if allowed to mix with oxidizing substances it will form highly flammable mixtures. Air is ubiquitous in most use environments and this leads to the primary issue, that of hydrogen's combustion behaviors in air. Mixtures of hydrogen in air are flammable over a very broad range and combustion is initiated by very low ignition energies such that any unplanned release may lead to fire and explosion hazards. The issues with physical behaviors, combustion, materials usage, health issues are highlighted.

6.5.1 *Physical Behaviors*

The following material introduces basic hydrogen properties, but also lays out how they can be a challenge to use safely.

Hydrogen States. To date, applications[1] have used hydrogen as a solid (SH_2), liquid (LH_2), slush (SLH_2), and a gas (GH_2). Hydrogen's physical properties arise from phase transition behaviors at very low temperatures and over a very narrow range:

- The triple point (where all three phases coexist) occurs at a temperature of 13.8 K and pressure 7.2 kPa [Safety note: This is a sub-atmospheric pressure, and it is critical that systems operating at such pressures *absolutely* prevent in-leakage of atmospheric air (or other oxidizing substance)],
- The normal boiling point (NBP) at an absolute pressure of 1 atmosphere (101.323 kPa) is 20.3 K,
- The critical temperature (the highest temperature at which hydrogen vapor can be liquefied) is approximately 33 K.

Atomic and Molecular Structure. Hydrogen, the most abundant element in the universe, might be consider a basic building block with its first-place entry in the periodic table. There are three known isotopes of hydrogen: protium (AMU 1); deuterium (AMU 2, one neutron); and tritium (AMU 3, two neutrons). The issues discussed in this text focus on molecular hydrogen formed from protium. Atomic hydrogen occurs by ionization or by dissociation of the covalent bond on a metal surface. Deuterium and tritium are constituents in nuclear processes with many specialized safety concerns beyond this treatment. However, if deuterium and tritium are concentrated, then their physical properties will vary according to their greater atomic weights, and because the chemistry of all three isotopes is very nearly identical the safety issues reviewed herein will apply. Only when sufficient quantities of deuterium and tritium are considered does radioactivity become a safety concern.

Concentrated as pure element, hydrogen under normal conditions forms a diatomic molecule with a covalent bond. The interior spin states of the two hydrogen nuclei in the molecule give rise to allotropic forms of hydrogen, ortho- and para-hydrogen. Ortho-hydrogen possesses parallel spin states while para-hydrogen has anti-parallel spin states. The ratio of the populations of molecules with the two spin states varies

---

[1]Transition of SH_2 into a metallic state occurs at pressures exceeding 200–300 GPa and may be superconducting.

with temperature such that higher temperature favors greater numbers of orthohydrogen and low temperatures, parahydrogen. This behavior, arising from the arrangement of nuclei within the molecule, has implications for the bulk properties of hydrogen.

General Gaseous Behaviors. Let's examine hydrogen's general physical properties at ambient temperatures to get a better appreciation of issues. To human senses hydrogen gas is invisible, tasteless, and has no odor. Aside from requirements that require high purity, adding odorants (like mercaptan used in natural gas) is of no value because hydrogen is so diffusively mobile that it will leave any odorant behind. Basic information is tabulated (Table 6.1).

Adding a further edge to these sensory deficit issues is the fact that the flames of this highly flammable gas in air are practically invisible under many ambient lighting conditions. Hydrogen flames emit in the UV and two bands in the mid-IR, but with relatively low amplitude in the visible range. And, there is yet one more sensory deficit twist! This arises from the low emissivity of hydrogen flames, except when large clouds of hydrogen are involved in combusting.

The emissivity is one to 10% of other hydrocarbon flames such that we can't sense a hydrogen flame's heat, meaning that we won't recoil from the heat gradient of the flame because we can't perceive it. Hydrogen flame temperatures in air are very high 2045 °C (3713 °F), so that without protection [fire resistant gloves] burns are unavoidable. There may be clues to the presence of hydrogen. Under certain lighting circumstances the density gradients formed by release of hydrogen in air or the heat from its combustion produce visible distortions, sometimes readily observed when light, incident upon the mixing of GH_2 in air falls on nearby surfaces. Leaking GH_2 may produce noise. With a sonic velocity nearly three times that of air, amounts of hydrogen mixed in air can alter ambient sounds, shifting their apparent frequency to the human ear (like helium). These basic physical realities present a real conceptual hurdle to new hydrogen handlers, namely if one can't rely on their senses to indicate where it is, how can **this** be safe! For starters, detection instruments *are required* for work with hydrogen.

Hydrogen does possess characteristics that can be used to advantage. At ambient temperatures hydrogen is the lightest of all gases and has a density a fifteenth that of air. This results in great buoyancy, and if it wasn't for the flammability, we'd all use

Table 6.1 Hydrogen general properties

Sensory properties	Colorless, odorless, tasteless
General properties	Flammable, non-irritating, nontoxic, asphyxiant non-corrosive
Highly buoyant	Lightest gas of all the elements and can escape earth's gravity (if ionized)
GH_2 density @ NTP	0.0838 kg/m^3 (1/15th air)
GH_2 specific gravity	0.0696 (air = 1.0)
Viscosity	33.64×10^{-3} kg/m h (1/2 that of air)
Diffusivity	1.697 m^2/h (4 times methane in air)
Thermal Conductivity	0.157 kcal/m h K (7 times that of air)

hydrogen in our party balloons! The rate of rise in air is remarkable at around 3 m/s under typical conditions, and even greater if the hydrogen is hot or the surroundings cooler. If the system in question is out in the open (outside) this is a boon to safety considerations. However, inside structures or other confining regions released hydrogen will seek the highest points, and where enclosure spaces are connected by lines or ducts will flow to the highest points in the connected regions by simple convection. [Several safety notes:

- If inadvertent release can lead to accumulations that can form flammable mixtures, electrical components (lights, switches, motors, etc.) are potential ignition points. Code requirements may specify explosion proof devices for these areas.
- Ventilation may be enhanced if accumulation point feed into vents.
- Detection instrumentation can be located to advantage in potential accumulation points.
- Electrical conduit in hydrogen use areas that convey control lines to control rooms or other unprotected chambers should be "potted" or sealed to prevent inadvertent hydrogen transport and accumulation].

The high diffusivity of hydrogen often is a benefit to safety. Remember, a concentration of gas, if not confined will always disperse until an equilibrium concentration is achieved. Hydrogen will do this better than any other gas.

Due to the risks of inadvertent release a high premium is placed on using high integrity seals to keep hydrogen in a system. However, hydrogen molecules are small, and in bulk as a gas or liquid possesses low viscosity giving handlers another concern; preventing it from leaking which it does so more readily than other fluids.

Even though at ambient temperatures hydrogen is the lightest of all gases, at cryogenic temperatures near NBP (see below) it can be slightly heavier than ambient air or neutrally buoyant.

General Cryogenic Behaviors. Hydrogen safety considerations vary upon whether cryogenic temperatures are part of the application. Significantly higher densities are achieved with the liquefaction of hydrogen, an important consideration for bulk storage, but in practice entails a whole new dimension of safety issues compared to those of ambient temperature GH_2. Greater densification is achieved by either reducing temperatures, producing 2-phase mixtures (SH_2), or by high pressurization of liquid or gas to 700 bar (10 kpsia) which achieves as much as a 30% densification.

LH_2 is colorless, has no odor, and possesses the lowest density of all liquefied gases (see Table 6.2). It is noncorrosive and is not considered highly reactive. The normal boiling point temperature (NBP) is 20.3 K (423.13 °F) and is less than the freezing point of all other gases, except for helium [Safety note: It is important to minimize impurities in cryogenic systems to avoid blockage in lines and components due to solidified gas impurities]. Hydrogen vapor is neutrally buoyant in air at 23 K. The robust buoyancy previously noted for hydrogen gas doesn't begin until vapor temperatures warm above cryogenic temperatures, defined at 123 K. Cryogenic

Table 6.2 General cryogenic hydrogen properties

Normal boiling point (NBP)	20.268 K (@ 1 atm/101.325 kPa)
Vapor density @ NBP	1.338 kg/m^3
Liquid density	70.78 kg/m^3
Neutral buoyancy temperature in air	23 K
LH_2 specific gravity @ NBP	0.0710 (H_2O = 1.0)
Equivalent vol gas @ NTP	845.1 (per vol liquid @ NBP)
Pressure to maintain NBP	172 MPa
Thermal expansion	0.0164 K^{-1}

hydrogen vapors released into the open will persist along the ground or elevation of release, and can be transported by wind[2] some distance before rising.

Exposure to ambient heat will result in the expansion of liquid hydrogen [Safety note: the coefficient of volume expansion is greater than that of water (23×). This has implications for the design of ullage volumes, relief equipment, and operations]. Direct exposure to ambient temperatures will cause rapid boiling, vaporization, and expansion as gas until at normal temperature and pressure (NTP) the total expansion over the original liquid volume will be a factor of 845 times greater [2] [Safety note: Safe handling of cryogenic hydrogen requires any hardware containing LH_2 to use redundant pressure relief protection to avoid over-pressurization, container rupture, and explosion hazards. This is a code requirement]. See also Chap. 3.

As a fluid hydrogen has a high heat capacity and coupled with the kinetics offers heat transfer characteristics that are useful in certain applications (cooling electric power transformers is one example). There is a curious detail that comes from spin transitions in the hydrogen nucleus. The heat capacities of ortho- and para-hydrogen differ, and at low temperatures para-hydrogen's heat capacity is significantly larger. This has implications for processing and storage of liquid hydrogen, as the inevitable self-catalysis of ortho-hydrogen to para-hydrogen at low temperature releases significant amounts of heat, capable of evaporating 1% of condensed product an hour. Therefore, catalysts are used to speed this process and remove this heat before it interferes with storage efficiency. Commercially supplied liquid hydrogen is para-hydrogen.

[Safety notes:

- LH_2 is **always** isolated from air or other oxidizers to avoid creating highly shock sensitive explosive mixtures formed by their condensation in LH_2.
- Uninsulated surfaces of LH_2 systems can condense air to form liquid air which presents its own hazards. These include frostbite and if liquid air combines with

[2] During tests (early 1990s) of the McDonald Douglas Delta Clipper, an effort to demonstrate vertical takeoff and landing by a rocket, hydrogen boiloff vented from the rocket crossed the New Mexico desert floor some several hundred feet through the launch exclusion zone to form explosive concentrations in support equipment housings. An air compressor was seen to blow off its cowling, making the appearance of a race into the sky with the rocket itself!

carbonaceous materials (grease, oily rags, plant matter, etc.) highly flammable and shock sensitive mixtures form.
- These low temperatures present a severe frostbite hazard to human tissues if direct exposure occurs. The danger of freezing by exposure to cold vapors is increased due to the high thermal conductivity of gaseous hydrogen.]

6.5.2 Combustion

Several criteria must be met before hydrogen can combust. It must be mixed with an oxidizer to form mixture within flammable limits, typically noted as the lower flammable limit (LFL) and the upper flammable limit (UFL). An ignition source must be present within the flammable mixture, however when the mixtures formed are near optimal, or stoichiometric, ignition is possible with such small sources of energy as to be considered spontaneous. This general statement applies to both gaseous mixtures of hydrogen and oxidizers as well as condensed phase mixtures which would be solid oxidizer mixed in liquid hydrogen.

Combustion of Gaseous Mixtures. Table 6.3 provides basic combustion data for gaseous hydrogen mixed in air. Flammable gaseous mixtures represent the most common hazard. As a point of reference, the minimum ignition energy (MIE) in a static electric discharge is several orders of magnitude below the human threshold of sensation.

Combustion can involve any one or all these processes: fire, subsonic flame acceleration known as deflagration, or supersonic flame propagation, specifically termed as detonation. Confinement of flammable mixtures is implicated in exacerbating hydrogen combustion events and is an important factor. If confinements are sufficiently narrow flame fronts lose enough energy such that combustion ceases, or the flame is quenched. The quenching gaps for hydrogen are narrower than those of other fuel gases.

Table 6.3 Gaseous combustion properties (para-hydrogen)

Flammability limits in NTP air	3.9–75.0 vol%
Flammability limits in NTP oxygen	3.9–95.8 vol%
Detonability limits in NTP air	18.3–59.0 vol%
Detonability limits in NTP oxygen	15–90 vol%
Minimum ignition energy in air	0.017 mJ
Autoignition temperature	858 K
Quenching gap in NTP air	0.064 cm
Diffusion coefficient in NTP air	0.61 cm^2/s
Laminar flame velocity	2.70 m/s
Estimated flame emissivity	0.01–0.10

Fire. Fire is a rapid chemical reaction that produces heat and light and is differentiated from deflagration and detonation by defining it as a stationary flame with the flammable mixture fed into the reaction zone (plume or jet). Fire is characterized by sustained burning and is accompanied by any or all of the following: light, flame, heat, and smoke.

Deflagration. Accelerated flames, or deflagration arise from a common hazard scenario where hydrogen is released within a system or an environment and allowed to mix with an oxidizer such as air to from a flammable mixture over a region. If ignited, flame will propagate through the flammable portions of the release, and typically moves at a subsonic rate with respect to the unburned mixture. Without confinement the flame advances by diffusion processes known as laminar burning which in air under ambient conditions produces propagation rates up to 3 m/s. When confinements have roughly equal dimensions (for example, roughly spherical or cube-like) the resulting pressurization from the combustion typically results in propagation speeds less than 100 m/s.

As a light gas the more rapid kinetics of hydrogen alter its behavior over that of other flammable gases. For example, its sound speed is nearly four times that of air and its diffusivity in air is four times greater than that of methane. The increased diffusivity has implications for the movement of free radicals in flame fronts resulting in higher laminar flame speeds and leading to coupling of combustion and fluid flow processes. This coupling promotes flame acceleration (deflagration, a subsonic process) in a bootstrap fashion. In confinements that focus acceleration along a "long" axis flame acceleration becomes significant. The rapid product gas expansion from combusting hydrogen-oxidizer mixtures readily interacts with any confining surfaces (rigid structures, pipe walls, and even the roughness on these surfaces, etc.) to produce pressurization, drive fluid flow and push the flame front into unburnt gas. In most circumstances the rate of reaction in the flame front is increased by greater mixing, releasing yet more energy (assuming the mixture has the necessary energy, >13% v/v) until further flame front acceleration is limited by choked flow, typically 400–800 m/s where the flame front is advancing relative to the unburned gasses.

Detonation. At this stage, should turbulent flow grow, shock formation can jump-start the deflagration to detonation transition (DDT) process[3] and result in a supersonic shock driven combustion, or detonation. In a sense, the energy available throughout a combusting mixture can be leveraged unpredictably against system elements downstream of the flame propagation. A more concrete, yet simple example, is provided to illustrate the possibilities:

Example

Consider ignition of a flammable hydrogen mixture in a steel tube. The confinement of the expanding product gases drives the flame front into the unburned mixture, increasing mixing and combustion and the rate of energy release. The

[3] A detonation is a shock combustion process that propagates supersonically (>1500 m/s).

flame propagates and as the hot gases expand and flow down the tube, they are subject to wall friction forcing the flame front to slow near the walls, but "nose" out toward the center of the tube, and in the process stretch the flame front. This increase in the flame front size improves mixing and further increases the rate of energy release. Incidental obstructions (bends in the path, obstacles, etc.) and surface roughness in the path of propagation induce turbulence in the flow that further increase the burning rate. While such interactions are common with a variety of fuel-oxidizer mixtures, with hydrogen they can lead to startling combustion behaviors. A sensitive[4] hydrogen-oxygen mixture ignited in a 0.5″ ID smooth walled steel tube can accelerate from a simple match ignition to transition to detonation within a *half meter* of run-up within the tube creating an overpressure roughly 18 times the initial fluid pressure and a jump in the rate of propagation to 1500 m/s and greater. There are other factors such as *pressure piling* (the advancing flame front pressurizes the upstream flammable mixture) and superposition of pressure waves (by reflection) that can lead in certain circumstances to combustion induced pressures nearly 1000 times greater than the initial fluid pressure.

This sort of possibility makes hydrogen system design a challenge, and if not properly accounted for, may lead to catastrophe. There are many factors to consider such as the mixture conditions (mixture composition, pressure, temperature, the presence of diluents), the detonation cell[5] size that provides kinetic information, as well as the confinement dimensions.

In general, simple combustion is more likely than flame acceleration, and flame acceleration is more likely than detonation, but with hydrogen-oxidizer mixtures these combustion proclivities are decidedly skewed towards the more reactive behaviors over that of other fuel-oxidizer combinations. With detonation, the supersonic propagation of the process defeats standard mechanical pressure relief devices which rely on subsonic fluid behaviors.

Condensed Phase Explosions. There are *uncommon* circumstances in which air or liquid oxygen can encounter liquid hydrogen. These circumstances result in extreme hazard, and while they are unlikely given standard hydrogen practice, they are mentioned specifically so they may be avoided.

The most exciting situation can occur with a rocket launches that fail during lift off, typically falling back to the launch pad, and oxidizer and fuel tankage collide,

---

[4]Well-mixed fuel and oxidizer components become more reactive and energetic as the mixture quantities approaches the optimum, or stoichiometric ratio. Such mixtures may be termed *sensitive*.

[5]Within the region of high compression of a detonation wave the superposition of intense acoustic waves and the shock front form combustion kernels or cells whose energy release drives the detonation shockwave. In detonation tube apparatus the high-pressure action of the cells against carbon sooted foils actually leaves diamond patterns etched in the soot (can appear as "snake skin"). The dimensions of the diamonds correlate with the energies that characterized the detonation, and whether detonation can propagate.

or a similar sort of occurrence of a violent failure on a rocket test stand. When liquid oxygen rapidly enters liquid hydrogen, it forms solid crystals and the motion of the crystals through the liquid hydrogen induces static discharge. The static discharge initiates combustion spontaneously and an explosion ensues involving the contents of the surface area of mixing. Typically launch accidents only involve a fraction of the inventory, but rocket launches do involve large quantities and the much greater density of the condensed phase over that of gaseous mixtures supports violent explosions.

A more pedestrian accident scenario, but potentially catastrophic is possible with what is called cryo-pumping. In this situation a leak passage in a cryogenic system exposes an internal super cold region to outside air. Liquid hydrogen is colder than all cryogens except liquid helium. The condensation of the air literally forms a vacuum in the passage and some of liquified air enters the hydrogen system. It freezes to form solid air. If this process proceeds undetected as a small steady instream a flammable mixture can build up in the hydrogen. Stoichiometric mixtures formed this way possess greater energy per unit mass than TNT and as a mixture approaches stoichiometry it becomes shock sensitive and can become more shock sensitive than nitroglycerin. A minor shock event, a jarring of a Dewar can initiate an explosion.

Finally, a liquid hydrogen leak onto the ground that persists for some time, can if conditions are right, end in an explosion. If the leak is sufficiently persistent to chill the ground to cryogenic temperature, water vapor will form a slush into which liquid air products along with some liquid hydrogen will accumulate until it becomes shock sensitive. At some point the falling drops of liquid hydrogen initiate a reaction.[6] This circumstance can only occur if the leak persists unattended. Industry experience is that most spills of liquid hydrogen rapidly flash to gas leaving a little frost behind.

If oxidizer and fuel are always isolated these situations will never arise.

6.5.3 *Hazards of Pressurized Systems*

Hazards inherent in hydrogen systems come from hydrogen's low heat of vaporization, its large liquid-to-gas expansion ratio and the large thermal difference between LH_2 and the ambient environment. At high pressure compressed gas possesses significant potential energy. Ultimately the cause of hazards originates with failure of containment components. Rupture and explosive release of hydrogen may arise from failure of the pressurization system, or failure of the pressure relief system.

Fire from an external source can introduce more heat into the system than anticipated in the design. With inadequate venting over pressurization results.

Before large scale commercial production of LH_2, users had to produce their own, and the issues of ortho- to parahydrogen conversion where not well understood.[7] Relief systems were overwhelmed by the larger than expected boil-off gas production.

[6] Observed by Health and Safety Laboratory, UK [8].

[7] This puzzled physicists until Heisenberg and Schroedinger's work with the development of quantum mechanics.

Operator error of overfilling creates an inadequate ullage volume. A swing in the internal temperature of the LH_2 can cause liquid expansion into a relief system designed only to accommodate gas. In such a situation the liquid-to-gas phase change creates large overpressures.

A Hypothetical. Theoretically, a scenario in which valves fail shut isolating liquid hydrogen in a line without pressure relief protection, otherwise known as "liquid-lockup", would result in pressure extremes as ambient thermal energy penetrated the thermal insulation. This may be examined two ways:

Consider the equivalent volume of gas at NTP per liquid volume at NBP: Ideal gas law predicts a gas to liquid volume of 845.1 times.
Alternative, one might evaluate what pressure is required at NTP to maintain liquid (NBP): 172 MPa (nearly 25,000 psia).

At some point, depending upon the strength of materials in the piping, the containment will fail at a weak point resulting in a rupture release of hydrogen. Not only is dangerous shrapnel possible, but a rupture event can create hot small metal shards capable of igniting a sensitive hydrogen-air mixture, should the two coincide, and this could result in flash fire.

6.5.4 Materials Considerations

Hydrogen systems incorporate materials for different functions, under a wide range of conditions and must be compatible with hydrogen's unique properties. Material selection may involve some of the following considerations:

- Design and operating conditions
- Effects of environment or operating conditions
- Change in physical properties over a wide range of temperature (inherent in cryogenic service)
- Corrosion resistance
- Toxicity
- Hydrogen embrittlement
- Cold embrittlement
- Thermal contraction
- Material cost, ease of acquisition, and availability of material's data
- Ease of fabrication, assembly, and inspection
- Behavior under fault conditions such as fire

The issues associated just mentioned are varied and complex and cannot be addressed comprehensively in this text. The recommendation is that designers consult materials specialist where critical conditions occur. Engineers and technicians working around hydrogen equipment must appreciate that where direct hydrogen exposure occurs selection of materials must involve critical review should substitution of materials be contemplated in a maintenance action or in the custom fabrication of assemblies. General concerns include:

Hydrogen attack on metals,
General material behaviors at cryogenic temperatures.

Hydrogen Attack or Embrittlement. Many metals and alloys undergo a decrease in fracture toughness or ductility when exposed to atomic hydrogen. Pure molecular hydrogen, and other hydrogen-containing gas species, particularly hydrogen sulfide (H_2S), hydrogen chloride (HCl) and hydrogen bromide (HBr) molecules. These molecules are adsorbed onto metal surfaces, and the hydrogen dissolves into solution with the metal atoms and diffuses throughout the material. Mechanical properties of tensile strength and ductility can be significantly reduced such that fracture loads may be reduces below the yield strength of the material. In some circumstances this process is rapid, occurring in minutes and in pressurized systems can directly lead to catastrophic failure.

There are three general mechanisms under which the effects of embrittlement occur with susceptible metals and alloys. Environmental embrittlement is seen in metals and alloys that undergo plastic deformation in a H_2 environment. The hydrogen enters through microcracks on surfaces exposed to hydrogen. For example, a vessel undergoing expansion at high pressure. The greatest effects are noted over the temperature range of 200–300 K. Internal embrittlement arises with hydrogen absorbed into the bulk of materials generally as the result of processes such as welding or electro polishing. Water present on work surfaces is electrolyzed, the hydrogen ionized and electrically driven into the metal. The history of components manufactured using such processes should be evaluated to ensure embrittlement is not a concern. This mechanism has its greatest effect with temperatures in the range of 200–300 K. The last mechanism of concern is H_2 reaction embrittlement that can occur within metals at elevated temperatures. The absorbed H_2 chemically combines with constituents such as carbon in steel to form methane or other brittle hydrides. The effect is reduced ductility.

The performance of several notched samples (Charpy Impact test) are tabulated (Table 6.4) to show the effect. Lower strength treatments generally perform better than higher strength counter parts.

There are a variety of factors that influence embrittlement behaviors including the operating environment, temperature, pressure, exposure time, and physical and mechanical properties. More detailed considerations may involve the stress state, stress concentrations, surface finish, microstructure, and existence of cracks. The purity and concentration of hydrogen must be considered. General strategies used with embrittlement issues are:

- Use less susceptible materials.
- Consider that the susceptibility to embrittlement generally increases with increasing tensile stress and alloy ultimate strength.
- Use increased material thickness.

Table 6.4 Some examples of embrittlement

Material (notched sample)	Exposure (at 80 °F)	Strength [MPa (psi)]	Change (%)
4140 (low strength)	69 MPa N_2 69 MPa H_2	1660 (241,000) 1407 (204,000)	−15.2
4140 (high strength)	69 MPa N_2 41 MPa H_2	2946 (362,000) 834 (121,000)	−66.6
C1025	69 MPa N_2 69 MPa H_2	730 (106,000) 552 (80,000)	−24.4
K Monel PH	69 MPa N_2 69 MPa H_2	1731 (251,000) 779 (113,000)	−55.0
K Monel (annealed)	69 MPa N_2 69 MPa H_2	993 (114,000) 724 (105,000)	−27.1

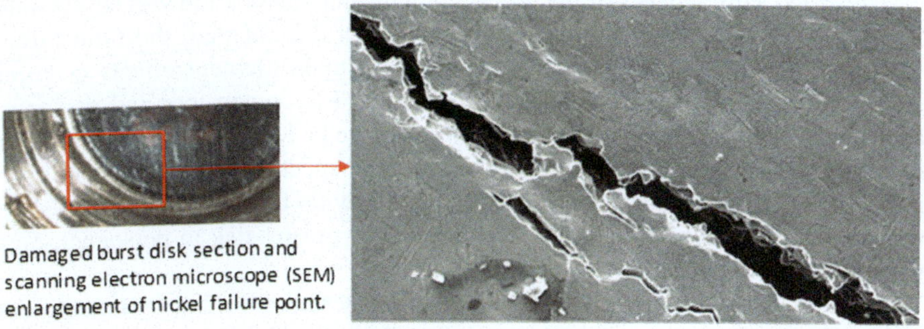

Fig. 6.4 Rupture disk fails when exposed to hydrogen [4]

- Reduce hydrogen exposure or purity.
- Account for processes such as electrical discharge machining (which may increase potential for H_2 embrittlement).
- Apply conservative design that accounts for the reduction in strength.
- Apply proper surface finish and welding (Fig. 6.4).

Example

A tube trailer was being converted from helium service to hydrogen service. The archived documentation on the system indicated the relief components were rated to 10,000 psia but failed to note a subsequent replacement of burst disks to nickel disks rather than the original 304 SS disks. The hardware was not physically checked with sufficient care to observe the change, and this would not be readily obvious. A leak check with nitrogen at 6000 psi was done without incident. Operations proceeded with charging the tube bank with the plan of reaching 6000 psia. Within approximately 20 min one of the burst disks failed at a gauge reading of 1600 psia. The operation was halted, gas purged, and a new burst disk installed. However, when the operation reached a pressure of 1800 psia a second burst disk failed. At this point all operation was suspended and an investigation of all burst disks was performed.

There are several points illustrated by this incident, beyond this is what hydrogen embrittlement looks like and that care must be observed for proper material selection. Knowing the pedigree of the tube trailer vessel steel (adequate for hydrogen service and had previously been used for hydrogen) would tilt expectation to an assumption that the hardware service corresponded and was selected for hydrogen use. The fact of successful helium service followed by testing with nitrogen at service pressure didn't provide a clue. A detailed check of installed hardware versus paperwork could have revealed a discrepancy, if the proper component identification was still legible. Should the service modification or repair had been documented, the future users would have been alerted to the change. Note the embrittlement effects occurred quickly, and that fortunately reduced the amount of hydrogen loss and personnel labor on the loading operation over what would have be lost if loading had gone on to yet higher pressures as intended. From a safety perspective, operations of this sort must account by planning for mishaps, because they do happen. And, also realize, that any unexpected high-pressure release entails a greater risk to personnel, even when proper precautions are being observed.

General Materials Service. In general, the materials specified by industry for ambient hydrogen service conditions are **not suitable** for cryogenic service. General concerns may be summarized as:

- Liquid air formation that occurs when components at cryogenic temperatures are exposed to air.
- Low temperature suitability of materials, including low-temperature embrittlement and loss of ductility of containment materials, differential thermal expansion of materials, especially at sealing junctions and considerations for materials in proximity to the low temperatures. See also Chap. 1.

Liquid Air Hazards. Uninsulated surfaces of a system containing LH_2 can readily condense air as shown in the photograph of an uninsulated chilled duct. It looks like water, but it is not harmless and must be treated with caution (Figs. 6.5 and 6.6).

Where a cryogenic stream collects, the exposed surface will be chilled to the point where air gases are condensed, and a slushy accumulation of water ice and condensed air gases will develop. Should liquid air come into direct contact with any carbonaceous materials as is typical in work sites (consider oily rags), flammable, even shock sensitive compounds form. When the liquefied air warms, nitrogen preferentially leaves the liquid mixture thereby increasing the concentration of oxygen from its starting concentration of 21% v/v as it exists in air to 50% v/v and greater. Oxygen enriched mixtures possess the same hazards as liquid oxygen!

Fig. 6.5 Liquid air formation [4]

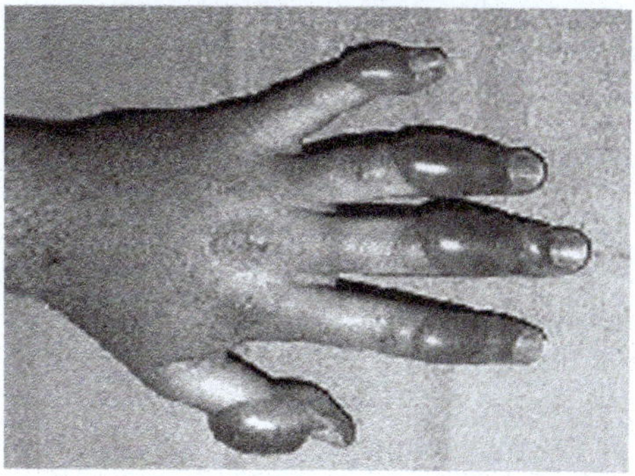

Fig. 6.6 Symptoms of acute LH_2 exposure [4]

6.5.5 *Health Hazards in Hydrogen Operations*

Extreme exposure to combustion or pressure hazards is avoided by system design or operational controls. Where hands on work occurs, hydrogen practice requires personnel use personnel protective equipment to mitigate minor exposures. In order of consideration these exposures are:

- Cryogenic exposure and hypothermia
- Thermal exposure
- High pressure gas or shrapnel impingement
- Shockwave exposure
- Asphyxiation

Cryogenic Exposure can cause symptoms like frostbite, and in more extreme exposures resemble severe burns. To avoid such consequences long sleeved insulating glover are used. In addition, coveralls are supplied without cuffs or pockets to capture liquid splash. Of equal concern is the danger that pressurized leaks of cryogenic hydrogen pose to skin and eyes. Protective glasses and face shields are employed for protection. Theoretically hypothermia is a concern, but this hazard would be difficult to realize in typical operations.

Fire. Because hydrogen flames are difficult to sense hands-on protocol will require personnel wear flame protective gloves and check hydrogen hardware for flames using fire detective instruments. Experienced personnel will apply situational awareness and look for indicators that forewarn of hazard; distortion of light, whistling sounds, knowledge of where leaks are possible, etc. The consequences for not doing so are severe thermal burns. Another combustion hazard is exposure to thermal radiation emitted by large flares, fireballs or UV radiation exposure in close proximity to hydrogen flash fires (but not direct exposure to hot product gases, mainly steam). Large fire balls or flaring of hydrogen produces thermal radiation similar to well characterized hydrocarbon fires and physiological effects may summarized as by noting a cumulative exposure of 2 calories/cm^2 will produce 2nd degree burns on exposed skin (Effects of Nuclear Weapons). More detailed information on exposure time and escape times is available (AIAA Guide). The effects of flash fire are not well characterized in the literature, but anecdotal accounts describe consequences like sunburn. However, it is important to note that hydrogen flash fires can occur within a second and deliver a total exposure capable to cause severe burns. Human reaction time is not sufficient to mitigate such consequences. Distance from the combustion is a critical factor because water vapor in the air can absorb the UV emissions (AIAA Guide).

Pressurized Components. Operations in proximity to pressurized systems are subject to code requirements. Proximity to pressurized components and fittings is avoided unless they are certified through pressure checks and inspection. Hands on activities require depressurization and removal of hydrogen. Hazards typical involve shrapnel or penetration of tissues by a high-pressure jet. See also Chap. 3.

Exposure to shockwaves (overpressure) over the body has the potential to cause significant physiological harm. Effects drawn from military experience are given in Table 6.5.

The protocols observed by standard hydrogen practice would isolate personnel from such effects. Exposure can only occur as the result of egregious error.

Inhalation. Hydrogen is nontoxic, but in quantity, exposure can induce suffocation by diluting the concentration of oxygen below levels necessary to support life. Asphyxiation hazards may potentially arise from hydrogen leaks, spills, or improper venting, and/or unplanned release of purge gases (He, N_2) all of which can result in an oxygen deficient atmosphere. This possibility is always a concern in confined spaces which are subject to occupational safety regulation but is also possible out in the open upon exposure to a large release. Situational awareness is critical because symptoms of physiological duress from lack of oxygen (12–19% v/v oxygen progress into loss of judgement at lower concentrations. Concentrations below 8% v/v oxygen

Table 6.5 Physiological effects of blast overpressures [4]

Maximum overpressure	Effect on personnel
7 kPa (1 psi)	Knock personnel down
35 kPa (5 psi)	Eardrum damage
100 kPa (15 psi)	Lung damage
240 kPa (35 psi)	Threshold for fatalities
345 kPa (50 psi)	50% fatalities
450 kPa (65 psi)	99% fatalities

are fatal within 6 min. Exposure to a highly oxygen deficient atmosphere can result in unconsciousness without warning! See also Chap. 4.

Protocols for hydrogen operation are designed to avoid directly exposing personnel to large amounts of hydrogen due to the fire hazard. However, hydrogen operations can involve inadvertent exposure to purge gases, typically helium or nitrogen. These gases are used to clear hydrogen systems of air or hydrogen depending on the need. Unfortunately, industry case history has many fatal accidents involving asphyxiation by purge gases.

6.5.6 Engineering Management

In organizations with a hierarchical structure such that a division of labors is in place, some consideration must be given to how to operate safely. In a facility scale environment all parties are responsible for safety. Overall responsibility and approval powers typically reside with management. In addition to oversight, management charter includes establishing the authority having jurisdiction (AHJ) for code compliance, managing pressure systems, hazard communications, providing personnel training, and establishing and insuring certification requirements for equipment and personnel are maintained. Conversely, personnel must be proactive regarding safety issues. It is critical that operations be provided adequately knowledgeable safety personnel with hydrogen experience (Chelton).

Other basic safety considerations include understanding physical properties in operating regime, incorporating failsafe design practice (redundancy, fail-safe operational characteristics), insure a "Safe Interface", that is critical operations and those involving personnel must be "two fault tolerant". Hazard Communications is based upon design and operation reviews, written and approved procedures, safety communications & training, mishap reporting, and the establishment of normative documents. See also Chap. 2.

6.6 Addressing Hydrogen Hazards

The methods of addressing hydrogen hazards pointed out in this discussion have their basis in aerospace practice and may differ in degree from industry practice. An excellent and comprehensive source of general practice information has been assembled by the U.S Department of Energy and can be found on-line at http://www.h2tools.org. In this section the general strategies and practices that form the basis of hydrogen safety practice are reviewed.

The assumption applied with this material is that the reader is working within a larger organization. Successful safety practice depends upon the decision makers and management exercising responsibility for establishing and enforcing safety policy. This includes ensuring that all applicable statutory and regulatory requirements are identified, documented, and followed.[8] In addition, management must define, designate, and document the entity, or entities, often referred to as the authority having jurisdiction[9] (AHJ) that is empowered to implement and enforce safety policies and procedures. However, personnel must also inform management as issues arise. Ultimately, everyone involved with H_2 system or operation is responsible for safety. To the degree necessary, support activities including procurement agents, janitorial staff, facilities maintenance, and anyone otherwise possibly exposed to hydrogen facilities or possibly impacting safe operation should be privy to hazard communications.

There are ample instances in hydrogen work where inadequate communications, a lax working environment and poor management control have led to catastrophe. A *culture* of written instruction required for safe handling/use of hydrogen. Written instructions or protocols:

- Should be formal (written),
- Approved and enforced by upper level management,
- Available to, and understood by, all personnel involved in H_2 activities,
- And, applied to all phases of system operations.

Personnel that work in hydrogen environments must have general hydrogen training and those who work with hydrogen systems should have specific certification for on the job training on those systems.

Earlier it was pointed out that safety begins with advanced planning so that the worst consequences are avoided. From a pragmatic approach, planners facing a particular issue will ask themselves several questions along the lines of:

- Can we ignore it?
- Is it possible to eliminate it?
- Is this something that might be avoided?
- Or, can it be controlled?

[8] Understanding and following regulation, codes and Standards, consensus standards and other guidelines is an integral part of operations, but specific information of this nature is not included in this guide due to the complexity of application in an international setting.

[9] The AHJ may be a person, a group, an office, an organization, or a federal, state, or local governing body.

To aid with these questions, and others like it, there is general guidance that can help. Plan to minimize consequences. If possible, site hydrogen work is performed separate from other activities. Limit as needed, the hydrogen work zones to essential personnel. Minimize the quantity of hydrogen involved. Design and operate for inherent safety. This can entail fail-safe design, for example incorporate valves that close upon loss of power. This strategy employed at the system level would lead to an automatic controlled shut down that isolates supply, removes hydrogen from the system and disposes of it safely. Caution and warning devices can advise personnel.

Design, safety, hazard, and operational reviews are applied to new systems as well as modifications to current systems. The results of the reviews are incorporated into approved operating procedures. All activities must include emergency plans that cover egress and response for all outcomes. Where critical oversite of processes is needed approved quality control and maintenance programs are put in place. If these safe principles and practices are used the consequences of operator error and equipment failure will be minimized.

How safe principles and practices are addressed is discussed in the context of:

- The notion of a "control volume"
- Basic Handling
- Systems
- Operations
- Hazards Assessment.

6.6.1 *The Notion of a "Control Volume"*

A hydrogen component, system, or facility might be considered a collection of confinements or volumes that possesses inherent design properties, sensors, a set of controls and is operated in a certain way to achieve whatever is intended. Figure 6.7 provides an illustration of the concept of a "control volume". Hypothetically, hydrogen is always contained in some sort of volume whether it is a storage vessel, a transfer line, or process elements such as an electrolysis stack, or a rocket thruster. The "controlled" and therefore "safe" hydrogen activity is not a function of one element in the system, f

For example a control circuit using input from a hydrogen detector but is a consequence of all the elements involved.

This concept may be applied sequentially from component to component or in a nested fashion such as a storage cylinder operated in a laboratory which itself is a control volume (has ventilation, detectors etc.). Even when hydrogen is purposely released to the environment it is done so under control and the environment becomes the control volume. The broader idea is for the hydrogen user to conceptualize their system in this manner.

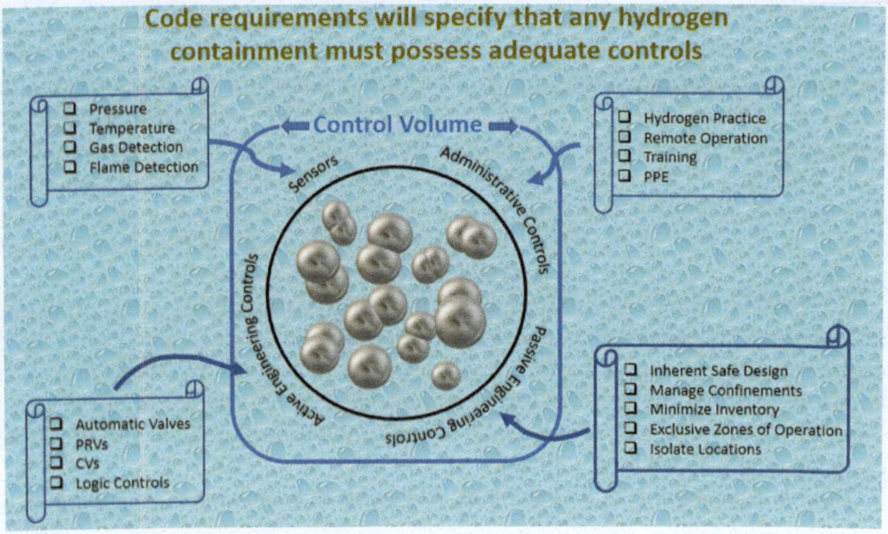

Fig. 6.7 Control volume concept illustrates the types of control elements that may be used to safely manage contained hydrogen [4]

Some of the benefits to this approach include:

- Understanding the interaction of system elements.
- It can aid in the identification and tracking of potential hazards. Instead of looking at a single fault in a component the effect on the control volume is assessed.
- Improving hazard analysis and hazard communication. A safety issue may arise from a leak from the system into a surrounding enclosed space. Rather than perceive this as a hydrogen system creating a hazard in the enclosure it is easy to make a case that the controls included with enclosure make the enclosure itself a control volume.

6.6.2 Basic Handling Considerations

Basic Handling Considerations. When operations with hydrogen are contemplated the overall goal is to achieve safe outcome regardless of what circumstances occur. This outcome is only achieved through design and careful planning of operations. Release of hydrogen, whether planned or unplanned must meet with adequate continencies. A simplified system is presented, and important implications of hydrogen releases and combustion are explored. Basic safety acumen is discussed as it pertains to components, subsystems, and common hydrogen activities.

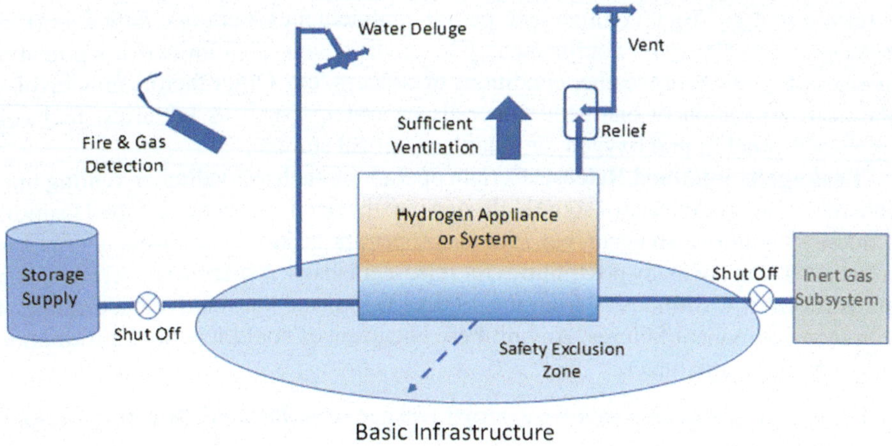

Fig. 6.8 Basic infrastructure elements in a hydrogen system or facility [4]

Basic Hydrogen Support Elements. What elements make up a system or facility? Aside from the primary hydrogen application, basic support elements are:

- A hydrogen supply which could be a hydrogen generator, a pressurized gas cylinder with manifold and regulator, or an external high-pressure tube bank or cryogenic liquid storage system,
- An inert gas subsystem,
- Ventilation and associated air monitoring,
- Relief systems,
- Vent disposal subsystems,
- Gas and fire detection, and
- Water spray or deluge.

Regardless of the application, the elements shown in the cartoon (Fig. 6.8) will likely play a role in operations. Other elements may be filters, vacuum appliances, sensors, and controls. In most facilities, hydrogen is supplied commercially over roadways either in steel cylinders, pressurized in tube trailers, or liquified in tankers. When hydrogen infrastructure becomes more common pipelines and delivery by ship may become common. The figure depicts outside fixed bulk liquid storage, but tanker trailers, tube trailers, and manifolded cylinders are all common place. Issues with these components and how hazards are addressed are examined.

General Combustion Issues. Past experience in working with hydrogen has identified combustion issues as by far the greatest concern followed by pressure system hazards, low temperature materials issues and personnel safety. At some point a system will invariably suffer an unplanned release. An understanding of potential hazards begins with identification of potential sources of hydrogen and where mixing with oxidizing substances might occur, knowing how much hydrogen is involved, whether it involves high or low pressures, and into what surroundings a release will go: Total confinement, partial confinement (openings, lattice work of beams or pipes?), or no confinement. The environment is examined for potential sources of ignition, or weather conditions of concern, etc. Other factors may involve the nearby location of energetic materials or systems such as batteries, fuel cell appliances, fuel, liquid oxygen, or volatile chemical storage.

Leakage/Unplanned Release. Certain operations such as loading or venting may inherently involve release of some hydrogen into the open, but in a controlled fashion. Undesired ignition can occur, but with the proper application of controls not present a hazard. There are many possibilities for release mishaps. Aside from operator error, statistics from U.S. industry from the 1970s [3] indicate the relative occurrence of common component failures. Any of these occurrences could lead to the formation of a flammable mixture.

- Leaks and spills either involve external leakage or in-leakage. Situations to look out for include:
  - Leakage between system components, especially when a component may harbor air and active electronics,
  - Secondary accumulation points such as the high points in a ceiling, or nearby rooms connected by conduit or vents,
  - Internal contamination such as air gases entering into a cryogenic system.

Within components there may be a variety of causes for leaks. For example, material causes may come from diffusion/permeation, expansion/contraction, embrittlement, low temperature embrittlement, corrosion, wear, or damage. Leaks can come from mechanical wear due to stress and vibration, deformation, pressure, or temperature. Unfortunately, an even greater cause is operator error 78). Internal contamination can happen with improper purging, introduction of contaminated fluids, pressurization gas, pump oils, and through a buildup of impurities that come in with each new supply of hydrogen over time.

Ignition. The safety strategy is to eliminate sources of ignition or keep them away from hydrogen. Enforcing an exclusion zone around a system is one step. It allows control over personnel and equipment they may inadvertently bring into a hydrogen work area. A sample list of items to be concerned with includes radios, phones and other mobile devices, generators, automobiles, etc. In the exclusion zone all electrical components need to be evaluated as presenting an ignition hazard. Critical components, defined as ones that either will be exposed to hydrogen, or potentially exposed to hydrogen will need to be explosion proof according to code requirements. Note that the term explosion proof refers to equipment with housing, seals and electrical features that will *prevent* whatever the function of the device *from igniting* a surrounding flammable gas mixture. The term has nothing to do with the "hardening" of the component to *withstand* an explosion. As an example, adequate lighting around hydrogen systems is a code requirement and explosion proof lighting fixtures may be needed.

Table 6.6 Possible ignition sources in common environments

Electrical	Mechanical	Thermal	Chemical
Static discharge	Mechanical impact	Open flame	Catalysts
Static electricity	Tensile rupture	Hot surface	Hydrides
Two-phase flow	Friction and Galling	Personnel smoking	Reactants
Static electricity (flow with solid particles)	Mechanical vibration	Welding	–
Electric arc	Metal fracture	Engine exhaust	–
Lightning	–	Resonance ignition	–
Charge accumulation	–	Explosive charge	–
Electrical charge generated by equipment operation	–	High-velocity jet heating	–
Electrical short circuits	–	Shock wave due to tank rupture	–
Electrical sparks	–	Fragment from bursting tank	–
Clothing (static electricity)	–	–	–

Hydrogen's low minimum ignition and broad flammability make releases susceptible to ignition. There are a multitude of possible ignition sources in common environments. A few are suggested by the tabulated possibilities (Table 6.6).

Highly sensitive hydrogen mixtures which occur near stoichiometry are so prone to ignition that combustion is characterized as spontaneous. Because of this it is recommended that hydrogen training countermand the common concept of the fire triangle taught about fuels in general.

Another aspect of releases is whether they occur between components within a system or constitute a release outside of the system and therefore can interact with the external environment. One such scenario is an external release of a large amount of hydrogen that is carried by the wind until it encounters an ignition source. The combustion then propagates to the source of the release, or "flashback".

Explosion. There are circumstances that will support explosive hydrogen behavior. The safety strategy is to avoid creating the conditions that can lead to overpressure, and if that isn't possible, keep personnel away from any such exposure. A summary of such circumstances includes:

- Allowing flammable hydrogen mixtures to accumulate in any confinement. Combustion within a confinement will result in overpressure.
- If the confinement resembles a pipe or duct in terms of relative dimensions flame acceleration is possible, and even higher overpressures are possible, and if conditions are right detonation processes may be supported.
- Situations that would permit mixing of liquid hydrogen and any oxidizing substance to form a condensed phase mixture. An example features an air leak into a line that is part of a cryogenic system. The air condenses and mixes with the hydrogen in a process called cryo-pumping (see item F in Figs. 6.9 and 6.10).

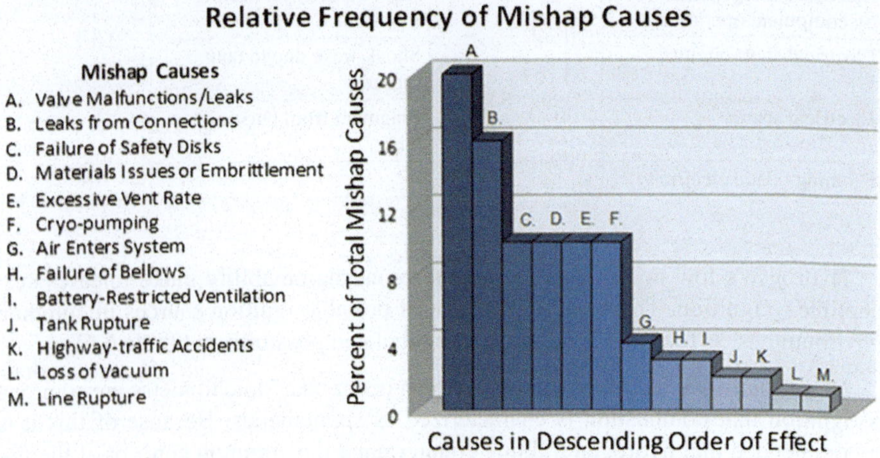

Fig. 6.9 The relative frequency of mishap causes as a percent of the total [3, 4]

Fig. 6.10 Simulated spill: 1500 gallons of LH_2 in 30 s at the NASA White Sands Test Facility in 1980 [7]. Condensed water vapor forms the visible cloud

6.6.3 Principles for Addressing Hazards

General. The design process is best served when safety is treated as an integral element and not as an afterthought. The general guidance in preventing leaks or unplanned releases and reducing the consequences when they do happen is to pursue the following general approach. Design and operations should incorporate attributes that work to prevent unwanted fuel/air mixtures. Once a flammable mixture forms it isn't possible to rely on suppression of ignition sources to maintain safety. Ideally such systems would always keep hydrogen and oxidizers apart from each other, until at the point of use. Using a purge process can help. Purging with an inert gas, typically nitrogen for ambient processes and helium for cryogenic systems is used to isolate hydrogen from oxidizers. Before hydrogen is brought into a system, air is removed using the purge gases. Air has been used as a purge gas, but only in circumstances where it can be done safely (small amounts of hydrogen, small diameter lines, small flames or overpressures do not pose danger, etc.). Before maintenance or repairs are initiated hydrogen is removed by purge from the system. Disposal of purge gases and hydrogen must be done safely. An industrial scale burn stack is used to combust large quantities of hydrogen by using a propane burner. Small releases may be released directly to the atmosphere above roofs or work areas as long as there is sufficient clearance, other exposures (electrical lines, etc.) are avoided and prevailing winds cannot drive hydrogen into ventilation ducts. The criteria of what defines a large quantity for disposal varies with jurisdiction, but NASA defines a threshold disposal rate [4] at 0.5 lbs/s (0.23 kg/s).

Minimize Leaks. Another goal is to have a leak free system. There is no perfect means to avoid leaks as seals and components will fail. In Fig. 6.9 it is clear that connections and valves are the biggest contributors to undesired releases. Therefore, a common-sense strategy is to avoid unnecessary use of valves and to reduce the number of connections to the least amount possible, then to weld as many connections as possible. The remaining valves and connections can be carefully monitored and maintained. The seals in a hydrogen system should be tested prior to putting hydrogen into the system. With gaseous systems a nonflammable gas is used to pressurize the system. To evaluate first for gross leaks, use nitrogen. Then helium, being a light gas and close in physical size to hydrogen is the best gas to use for high pressure tests. If there are small leak passages helium will reveal them better than other gases, nitrogen for example. Helium is expensive, test gas combinations, or equivalent leak measurements have been worked out (DOT).

Permitted Hydrogen Exposure. Because hydrogen is flammable well below the concentrations that would threaten asphyxiation, ventilation and controls are required. A space with limited ventilation is considered a confined space, and there are specific code requirements for the safety of personnel working in those environments. The amount of hydrogen permitted in a confined space environment is usually set at a tenth of the lower flammability limit (North American standards). In standard occupancies the measure of safety is to keep the amount of hydrogen in an air environment below a ¼% by volume. A different but related concern is keeping hydrogen or air, depending on the circumstances, out of an enclosed space. This may be done by maintaining positive pressurization. Liquid hydrogen storage vessels are managed not at 20 K, but at slightly elevated temperatures so that there is a pressurized ullage. A leak, by proactive design, will be a leak of hydrogen into air rather than the reverse which could pose greater hazards. Alternatively, control rooms that might suffer temporary and limited exposure to a hydrogen cloud, are pressurized with air to keep hydrogen out.

Operations. Operations are planned so that personnel are excluded from any situation, and especially circumstances that can occur due to equipment failure. Work is performed with hydrogen removed. Exposure is limited to hazards that can be mitigated with personnel protective equipment, otherwise remote operations are relied upon. Equipment operates according to fail-safe design and redundancy is employed in all critical areas. Exclusion zones are controlled to keep non-essential personnel away from hydrogen operations and storage areas, and to kept personnel out when hazardous operations commence. Examples of routine hazardous operations include:

- Receiving and transfer of hydrogen
- Disposal operations that involve large quantities

6.6.4 Components

The basic elements that support the hydrogen system or process include the hydrogen supply, connecting lines, relief components, inert gas purge, the disposal subsystem, fire and gas detection. Issues for basic components and subsystems and how they might be addressed are discussed by topic.

Joints and Connections. Welded connections are superior to other connection methods and is recommended wherever possible. Threaded connections are discouraged, but using proper thread sealant, may be used with GH_2. However, they are not used for cryogenic service where the sealant will solidify, shrink and crack. Soft-solder joints not permitted because of their low melting point. A small leak if ignited will melt out the solder and result in a larger release. Demountable joints, such as flanges, should be used only when necessary for junctions that need assembly, installation, or maintenance. Bayonet fittings are used for demountable LH_2 connections.

Valves. Valves are used in a variety of important functions. Isolation valves separate a component, typically storage from the remainder of a system. No additional valve is permitted between an isolation valve and the component protected. It is required for emergency and maintenance. Emergency isolation valves are used for manual or automatic shutoff at source. Other functions:

- Excess flow valves are used to isolate from a downstream equipment failure,
- Check valves prevent backflow and prevent contamination. In aerospace hydrogen service poppet types are used with line sizes ≤ 0.38 in. and swing or lift type for

line sizes 0.5 in. When bubble free tightness of components is needed check valves should not be used.

The valve bodies and soft goods must be compatible with the operating conditions. For GH_2 service, most general industry valves are acceptable. Remote actuated valves may be actuated pneumatically or by electric solenoids. Typically. this mechanical function is located directly over the packing such that a hydrogen leak will make contact. Because solenoid valves are an electric component, without applying some mitigation any leak would be readily ignited. If the solenoid valves are sealed. then they may be acceptable for hydrogen service. For LH_2 service, cryogenic globe or globe-type valves are recommended, but plug or ball-type may be used. The ball valve design should preclude trapping liquid in the ball when the valve is closed to avoid a liquid lock-up hazard. Shutoff valves shall not be installed between a relief device and a volume being protected.

Pressure Relief Devices. Relief devices are required for any volume in which LH_2 or cold GH_2 could be trapped and not have an escape path to relieve pressurization should the cryogen warm. Two common designs are a spring-loaded relief mechanism that will close once the pressure has dropped below a set point and a burst disk, designed to rupture open rapidly and permanently when pressure exceeds a certain level for a enough time. Relief devices, including soft goods, must be selected for the conditions of operation. The metal used in burst disk must not be embrittled by hydrogen. Relief devices should be set to limit pressure to maximum allowable working pressure (MAWP) for the volume in question. Given the high possible pressures which if not relieved could result in an explosive rupture, redundancy of relief devices is commonly required, and even redundancy in types (relief valves and rupture disks) is preferred. This practice prevents a single condition (an example would be ice formation) from causing the failure of both devices. A maintenance schedule is required to keep relief systems operating properly. Several notes apply:

- Rupture disks must be routinely replaced due to their finite lifetime,
- Cyclic loading can accelerate the failure of rupture disks. To accommodate this the rated burst pressure should be selected sufficiently above the intended operating pressure to avoid premature failure.

Hydrogen supply pressures are often greater than the MAWP of connected components. Therefore, downstream system elements supplied by a regulator connected to a high-pressure source should be protected by a relief device unless designed for the maximum pressure of the source. Another system element to which this caution might be applied are vacuum systems. Should a cryo-pumping occur with a vacuum jacket a considerable amount of air gases can freeze on cold inner surfaces. If the system is shutdown, and this condition isn't detected, the air gases can rapidly expand and rupture the vacuum jacket. Therefore, vacuum volumes should be protected with relief devices as depicted in Fig. 6.11.

The flow capacity of a relief devices must be sized to cover all phases associated with an operation. Relief device outlets should not impinge on other components or personnel or be manifolded unless pressure effects can be determined. Relief devices

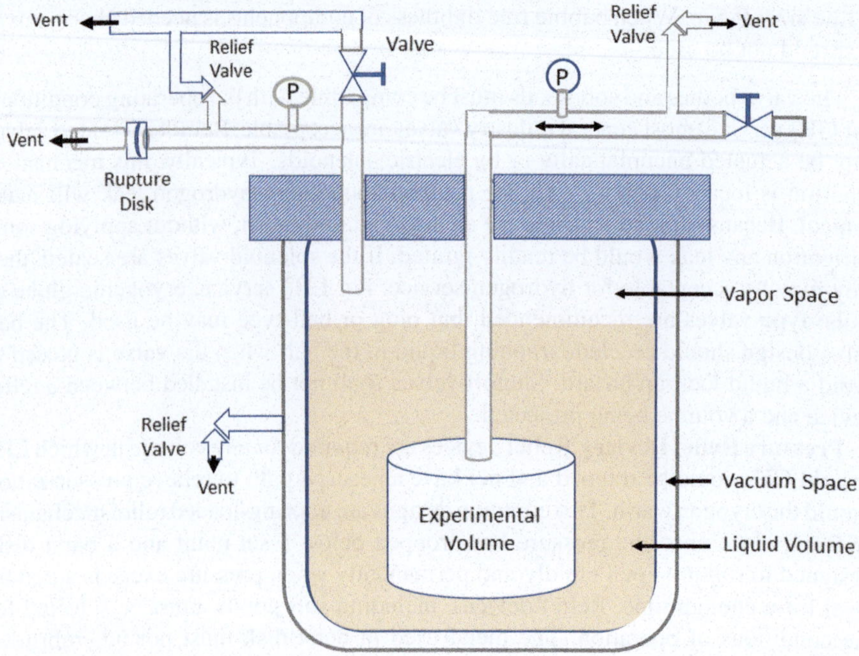

Fig. 6.11 Illustration of relief protection for typical volumes including liquid storage, interior cryo-cooled volumes and a vacuum annulus [4]

should discharge to a properly designed vent and should not discharge where H_2 can accumulate. See also Chap. 3.

Instrumentation and Controls. Modern hydrogen systems use a variety of instrumentation and controls. Consider instrumentation provides way to obtain quantitative measurements of behavior or state of a process while controls provide a way to maintain or change behavior or state of a process (for example, the thermometer vs. thermostat analogy). Any volume containing H_2 should have adequate instrumentation and controls to ensure that operation is within acceptable limits. Given the desirability for inherently safe systems the system must be able to adapt to different operational conditions, even faulted conditions. Typical instrumentation uses include:

- Instrumentation and controls as necessary for automatic operation,
- Minimally adequate instrumentation and controls to monitor and control its operation,
- Provide performance data,
- Provide warnings and alarms for out-of-limits conditions,
- Provide adequate notice when a hazardous condition is indicated.

Hydrogen Detectors. Recall that unaided human senses have difficulty detecting gaseous H_2. It is colorless and odorless, and while operators can be attuned for hearing an unaccountable sound that may be a leak, the leak sounds may be masked by background noises. The wide range of flammability and potential for fire initiation by static discharge from personnel makes this a critical concern that drives a need for caution. Unfortunately, detection isn't a case of one type suits all occasions.

A variety of factors drive selection of a detector that will suit particular needs. The goal is to provide immediate detection and warning of leakage. Where detection of flammable mixtures is important the detection must be able to indicate at least 1% by volume in air (25% of LFL) and provide signals to enable shutdown at 2%. This is not required on outdoor locations. Consideration of the use requirements, detector characteristics and limitations are necessary. Primary considerations are:

- Response time,
- Minimum gas concentration detection requirements,
- Full-scale range of the detector system,
- Level of concentration for which alarm detection is required,
- Accuracy of sensors,
- Reliability and recalibration frequency,
- Interface to facility safety and shutdown systems,
- Physical interferences, such as water condensation, soot, grease that may affect the sensor's ability to function?

More detailed factors include the effects of gaseous contaminants, interference against which the system can't discriminate (for example, a catalyst system may not be able to distinguish volatile vapors that can be present from hydrogen), and poisons that may compromise sensing mechanisms within the detection system. It is important to understand whether oxygen is required for the sensing mechanism to work (for example, catalyst-based sensors need an oxidizer present). For instance, an oxygen-based sensor may not work in a vacuum. Deployment issues may be important if environmental effects such as wind, buoyancy, etc. can affect performance, and even more critical is if the sensor act as an ignition source, a generally undesirable characteristic. There are different types of H-detectors technologies:

- Catalytic,
- Electrochemical,
- Semiconducting oxide,
- Thermal conductivity,
- Mass spectrometer,
- Sonic,
- Optical,
- Glow plugs.

It is common practice to locate fixed detectors at likely leak locations such as valves or flanged joints. Portable detectors should be used by personnel wherever H_2 might leak or accumulate. Keep in mind, that where large inventories of hydrogen are concerned that the overall status of the system, or facility must be consulted before

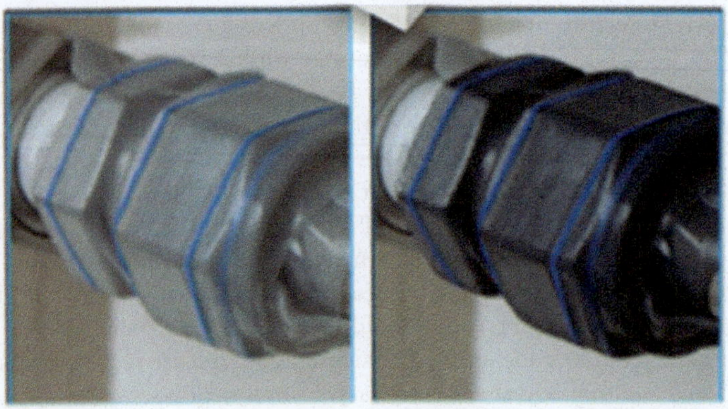

Fig. 6.12 Information from element one, Boulder CO., USA (*Disclaimer* At this time, this product is unique and worthy of mention (endorsement is not intended) [4]

entering in proximity. The forward path indicated as clear by a portable detector will give no warning of a large eddy positioned to engulf personnel.

There are passive detection tools that have valuable capabilities. Chemochromic silicone tape will change color on exposure. It has the characteristic of "a saran wrap like plastic" and can stretch fit over components and the overwrap will stick to itself without adhesives. The color change visible from a distance.

Fire Detection. Again, human senses are inadequate, and operations must rely on instruments because hydrogen flames nearly invisible in day and the emissivity of H_2 flame is so low that it is difficult to sense radiated heat. It is not uncommon in complex piping arrays for a hydrogen fire to be discovered that must have been burning for some time. There are a variety of detector types, and they work by different technology:

- Temperature sensing
- Heat sensitive cable
- Optical
- Broadband imaging
- Narrowband imaging, for example UV detection
- In proximity to the flame a straw broom or tossed dust
- Fire extinguisher (Fig. 6.12).

During the last decade multiband detectors have become common. They avoid inadvertent triggering of UV detectors which easily register false positives to reflected sunlight, welding and lightning. The multiband detectors require line of sight and the response time and field of view can vary with range. With proper fixed positioning they can monitor a considerable region. Another option for improving surveillance is to mount them with a video camera on a motorized gimble for remote operation.

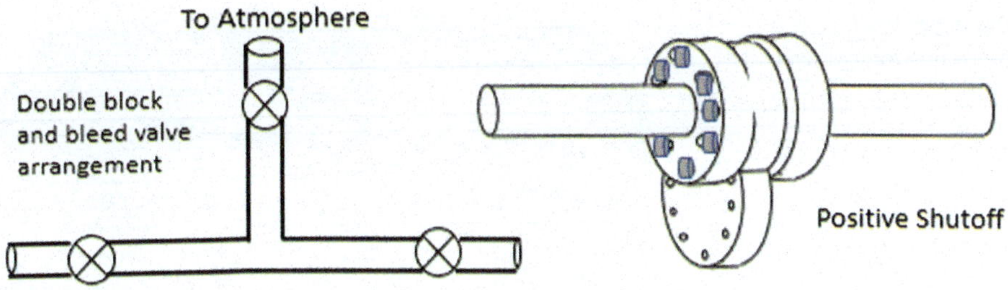

Fig. 6.13 Means of positive isolation [4]

Personnel should use portable detectors where remote monitoring isn't applicable. The effects of lightning, sunlight, welding on UV/IR detectors should be understood. Also, detectors must not be an ignition source.

Filters. Contaminants and particulate in a system can block passages. In systems with high pressure particulate can be accelerated to high velocities and cause damage to internal lines and vessels. Filters can help solve such issues. They are useful in refill or resupply lines. However, they introduce a pressure drops that interfere with system function. Therefore, use the right quantity and locations to minimize system impurities. It is important to consider operability and seal/seat leakage. The design layout should locate filters where they are accessible for cleaning. Sintered metal elements have been known to shed particulate and should be evaluated to ensure this is not a problem. Non-calendered woven wire mesh filter elements do not have this issue.

Inert Gas Subsystem. The goal is to always separate hydrogen from air. This requires an inert gas subsystem with pressurized purge gases, typically nitrogen for applications above 80 K, and helium where there are cryogenic operations (GN_2 will freeze). A hydrogen system should be capable of being purged and vented. The connections arrangement to the primary system should protect the inert gas subsystem from contamination by backflow of hydrogen. This can be managed by supplying the purge gases at a higher pressure, using check valves, or providing a double block-and-bleed arrangement (a "tee" configuration with 3 valves) (Fig. 6.13).

There are several purge techniques that may be used:

Evacuation and backfill requires a vacuum pump. The vessel in question must be capable of withstanding sub ambient pressures without collapse. An intermediate fill with nitrogen can accelerate the process and dry the inside of the vessel.

Pressurization and venting require repeated pressurization followed by venting. At least three cycles are required to reduce hydrogen concentrations below the flammability limit.

The "flow-through" technique while simple poses a concern when dead head spaces (Figure pp) and complex geometries exist in the system.

A caution is mentioned against simplifying a purge by isolating a system into regions. When evacuation and backfill is used the fluid being removed may leak back into the evacuated region from an adjacent separated volume. Explosions have

Fig. 6.14 Personnel protection for hydrogen [4]

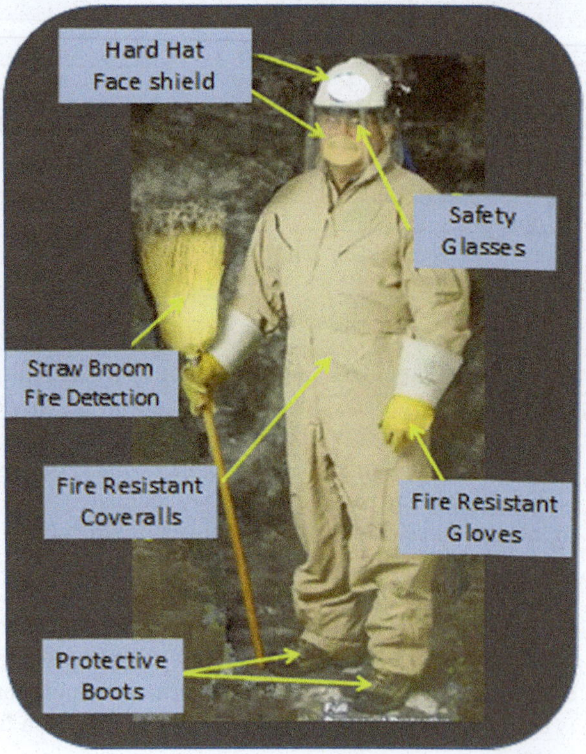

Fig. 6.15 Dead headed space illustrated [4]

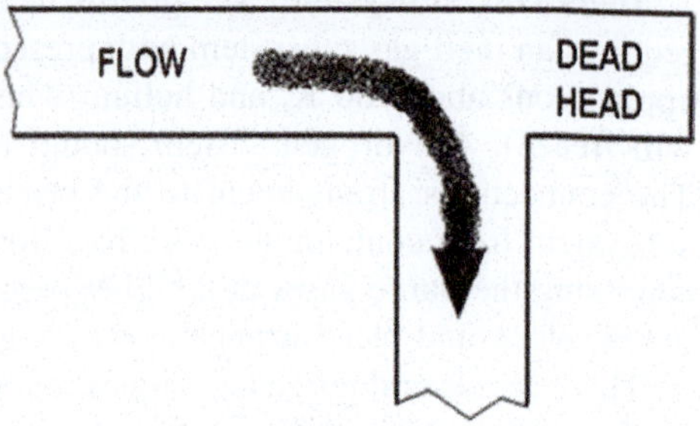

resulted. In addition, purge gases are often used in venting operations as a means to quench inadvertent burning of vented gas (Figs. 6.14 and 6.15).

Personnel Protection (PPE). Personnel Protective Equipment (PPE) is used to protect personnel from *limited* exposures to cryogenic or flame temperatures. The equipment is warranted for safety issues of exposure to cold surfaces or small quantities of liquid cryogen, oxygen-deficient atmospheres of inert purge gases (N_2, He), small pressurized jets of hot or cold gases, thermal radiation from H_2 fire, including intentionally flared H_2, and direct contact with inadvertently undetected hydrogen flames.

To protect personnel insulating cold surfaces where personnel are required to work and containment of liquid air with catch pans is recommended. Then it is up to operations to ensure personnel wear protective equipment to minimize injury where exposure is possible. Examples of conditions and PPE include:

- Operations involving cryogenic fluids require eye and hand protection, for example *face shields* and *cold protective* gloves,
- When connecting and disconnecting lines/components a *face shield* is appropriate to protect against emissions, pressurized splatter, etc.
- *Insulative coveralls of cotton/Nomex material* to be fire resistant. Where cryogenic fluids are involved the coveralls should avoid pockets and cuffs to not inadvertently catch liquid. Note, that fire resistant coating on fabrics wash off after a certain number of uses and must be discarded and replaced.
- Against small undetected flames, *fire resistant* gloves, *safety glasses* and *face shield* are used.
- Feet should be completely enclosed and protect by *closed-toe* shoes or boots,
- Aside from a generally noisy environment, ignition of a small amount of mixture can result in a loud pop, therefore, *hearing protection* as appropriate,
- Aside from the usual hazard in an industrial-like setting, liquid air dripping from elevated locations may be carried by wind gusts. Therefore, hard hats as appropriate.
- Portable detection to detect hydrogen gas or fire and to detect oxygen deficiency. A sweep of the broom is effective at safely detecting hydrogen flames over suspect areas (valves, connections, etc.).

6.7 System Considerations

6.7.1 System Considerations

Primary and necessary system attributes and considerations in include its location, failsafe operation, redundancy of critical components, proper placement of sensors and controls. Safety issues are reviewed by topic.

Location. How and where a hydrogen system is located, or how adjacent operations are managed are primary considerations. From the perspective of codes, the best place for a hydrogen system is outside, unencumbered by any enclosing structure, and separated from other activities and exposures. Of course, this isn't practical for most hydrogen applications. However, practicality of location is usually not the governing criteria in where a system may be installed. The inventory, in-process quantities, type of use and disposal requirements will feed into code assessment of a safe "footprint". Once the required exclusion zone is understood then search for a location can begin. Considerations will then advance to practicality and issues from pre-existing infrastructure that would interfere with hydrogen system operations. Because of this reality initial planning should ascertain what minimum quantity of hydrogen is necessary at each stage of an operation. Planning should consider minimizing storage, reducing transport quantities and the routes transport will take, minimize transfer quantities in delivery, as well as the end-use quantity. This approach reduces siting and area control requirements, and ultimately will mitigate the consequences of accidents.

Failsafe Operation. Controls the system to a safe mode upon detection of out-of-limit conditions of system functions or the operating environment, component failure or power loss. This can be supported by components that assume a safe state, for example, a valve whose unpowered state is closed. Or, by a control algorithm that move a system to a safe state. It is also accomplished by using redundancy for critical functions like pressure relief, isolation, and detection.

Use Alarms & Warning Devices. Warning devices should provide an alarm for potentially hazardous situation, preferably before it happens. Typical conditions that are monitored are abnormal conditions, malfunctions, and incipient failures. Alarms can flag control systems and can be audible, visible, or both. Some examples are:

- Pressure extremes
- Hydrogen in building ventilation intake
- Flare flameout
- Loss of vacuum insulation
- Valve position
- Pump speed extremes
- Hydrogen leak
- Filter differential pressure
- Fire.

Storage Vessels. From the perspective of regulatory code, the storage system is defined as including all the hardware and components up to the inlet of the hydrogen consuming application. For all but small applications that can operate with the output from hydrogen generation by electrolysis, hydrogen must be procured from commercial gas suppliers. The equipment used to store hydrogen can be systems unto themselves. They range from a single cylinder of gas with a valve, to manifolded groups of cylinders, to tube banks and finally to bulk storage that uses cryogenic storage vessels that may hold anywhere from several thousand gallons to more than one million gallons of liquid hydrogen. The scale of an application will determine what kind of storage is required. Laboratory operations and small-scale processes typically use gaseous hydrogen that is transported by truck to the point of use in carbon steel cylinders. These cylinders meet code requirements for highway transport and typically hold a kilogram of hydrogen under high pressure. Larger scale operations can make use high pressure gas tube banks that are mounted on a trailer suitable for highway travel. Cryogenic bulk storage systems are essentially large thermos bottles that are periodically filled. And, like a thermos bottle if the contents aren't consumed, they will eventually warm. For liquid hydrogen, even with very effective insulation the heat leak into the system is responsible constantly vaporizing liquid to vapor. This reality has implications that complicate operations. When a liquid hydrogen is first introduced to a vessel the inner containment is warm, so the liquid hydrogen flashes completely producing copious quantities of gas until the inner vessel is chilled to operational temperature. The pressurization and relief subsystems must be able to manage the inner pressure of the vessel over a wide range of temperature conditions. Recall hydrogen's liquid physical property for a large volume expansion with change in temperature (23 × water). Design requires operation must always keep a certain minimum gas volume, otherwise known as the ullage space, in the vessel. This permits a degree of thermal swing in the system before the liquid specified as the rated capacity would expand into the relief system [Safety note: this is a very dangerous situation that can cause the relief system to fail]. The relief mechanism must handle the great range in vaporization levels without exceeding the systems MAWP. Unless replenished unused liquid will boil away. Advanced insulation techniques and liquid helium chilling systems are being investigated to reduce boiloff and to densify liquid hydrogen in storage [5].

A brief accounting of controls is given. All storage systems are equipped with shutoff valve. Gaseous systems include a regulator, lines that convey the hydrogen to its point of use, relief protection and some means to remove the hydrogen from the line through a vent. An inert gas system may be used. A vaporizer is used to flash liquid where gas is required for operation. The gas is typically accumulated at required pressures in an interim high-pressure gas storage vessel. Generally, controls for automatic operation are used and are located for personnel access near the ground. The storage systems must use approved vent and pressure relief systems. Bulk storage system layouts must provide barriers to protect against the potential failure of rotating equipment, such as pumps, or from vehicles. In cryogenic systems insulation is managed by vacuum jacketing, heat reflective insulating materials and coating, and aerogel insulations. Foam insulations should be self-extinguishing (and not pose fire and explosion hazards when exposed to liquid air). After numerous loadings the trace frozen contaminants residing in liquid hydrogen can accumulate. The system may require periodically warming to remove these solid contaminants. Storage systems are electrically bonded at all joints and grounded to code requirements. Vessels must display their contents, capacity, and MAWP according to code requirements. Fluid direction in the lines are required by code to be labeled. Vessels is required to surrounded by a 15-ft clear space. When large quantities of liquid hydrogen are stored in proximity to other operations the base of the installation will include a catchment to contain a spill. This containment typically contains a bed of rocks into which the liquid would flow whose thermal mass will accelerate vaporization and aid dispersion of the hydrogen.

Safety consideration must consider all of the components in a storage system. A word of caution, where a third party or gas supplier provides the storage system as well as the product hydrogen, care must be taken to properly integrate safety concerns to include both the vendor supplied system and the system which it serves.

Piping Systems. The piping subsystem includes pipe, tubing, flanges, bolting, gaskets, valves, relief valves, fittings, and the pressure-containing portions of other piping components. Other elements included in a piping system are hangers and supports and hardware necessary to prevent overstressing pressure-containing components. The unique properties of hydrogen can require that special consideration be applied in the design and construction. Design considerations not only include pressure and temperature but must include the various forces that might impact the design as in a cryogenic system. The design must account for the most severe conditions. Piping of cryogenic fluids requires appropriate flexibility in expansion joints, loops, and offsets. Supports, guides, and anchors may have to account for expansion driven movements in the piping. The piping system must be electrically bonded across all joints and grounded. Code requirements must be applied to where piping is located including labeling (contents, flow direction). Concerns are that piping not located beneath electric power lines and that it be protected from potential failure of rotating equipment and from vehicles. Burying piping is too be discouraged, and if absolutely necessary, special considerations must be applied. Alternatively, subgrade lines can be placed in trenches covered by removable grating.

Applicable U.S. Codes and Standards

- 29CFR1910.103,
- ASME B31.12, ASME B31.3,
- NFPA 55 [Supersedes NFPA 50 A and B],
- CGA G-5.4.

Disposal Subsystems. Discharge is piped to a properly designed vent stack, that is outdoors, directed upward, and not impinging on structures, intakes or personnel. Small venting rates, less than 0.5 lb/s may be discharged as a gas. The discharge point should be located at a safe distance, above surrounding structures. Dispose of large quantities of H_2 by flaring, an operation that requires a burner and consideration for the amount of thermal radiation that will fall on surroundings. Vents shall not discharge where H_2 can accumulate. The discharge point of the vent must be protected from collecting moisture or in some circumstance the intrusion of insect or other nests. In most circumstances a purge capability is needed, therefore access to an inert gas supply is a consideration. The vent system design should account for H_2 vent velocity. Supersonic vent velocities can pose a noise hazard and require that piping be adequately mechanically bolstered against loads created by the expelled gas.

Air entrained in a vent system can cause flammable mixture formation. Ignition by static charge or electrical storm potentials can result in a gas vent becoming a flare. If the flammable mixture forms over some extant in the vent piping accelerated flames and overpressures can occur. This is undesirable, but usually unavoidable. Vent designs must account for such possibilities. Where accelerated flames can't be tolerated flame arrestors may be used. Vent or flare according to approved methods.

One such method is to use either a molecular seal or flapper to prevent air and precipitation from entering vent/flare system.

A design evaluation is required to ensure relief device connection to manifold does not affect relief pressure (Figs. 6.16 and 6.17).

Cryogenic Systems. Hydrogen systems that work at cryogenic temperatures are normally insulated with vacuum jacketed lines and foam to reduce heat input and boiloff. They also prevent liquid air formation and cold surface contact hazard for personnel. Where vacuum jacketing is not practical foam or aerogel insulations may be custom fitted to the element in question. Where flexible coupling is required

Fig. 6.16 Hydrogen vent system with propane burner [4]

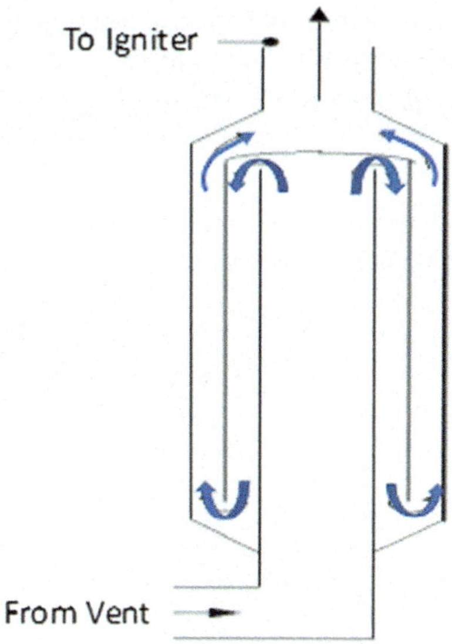

Fig. 6.17 Illustration of "molecular" seal design for a flare stack inlet [4]

metal convoluted flexible hoses and bayonet couplings are used. Ice build-up can be tolerated where component function is not impeded. Ice itself acts as an insulator. However, if ice build-up is acceptable personnel must be advised to leave the ice alone and not try to remove it, and not by "banging"!

Other notes and recommendations are:

- To use appropriate seals for cryogenic systems. Do not use thread sealant in LH_2 systems. When placing system elements containing seals back in service, after an initial ambient temperature leak check, "cold shock" and retighten lines and fittings. Then retest leak tightness.
- Both normal flow and cooldown actions need protection suited to their specific needs.
- Sudden pressure decrease on relief valve actuation will cause sudden boiling.
- Avoid thermal cycling on rupture discs which may induce premature failure.
- Moisture collected on or in relief valve can freeze and prevent valve from operating.
- Cracks or openings that expose cold surfaces to air can result in air liquefaction. Certain foams [Safety note: Styrofoam can be explosive if exposed to liquid air] should not be used as an insulation material. Insulating materials should have self-extinguishing fire rating. Factors to evaluate are whether the foam is an open cell or closed cell type, the cell size, what interstitial gas can form, and the effects of joints and gaps (Fig. 6.18).

Vents serving cryogenic systems must be sized to allow for flow under all conditions of operation, including quiescent boiloff, normal flow and cool down. Vents should be at least rated for 150 psig per CGA G5-5. Precautions must be taken to prevent cryopumping and moisture collection in a cryogenic vent system.

Vacuum Subsystems. Incorporated in applications to maintain insulating vacuum or remove unwanted H_2 or other gases as part of a purge process. Vacuum pump exhaust may have concentrated hydrogen, so it must be connected to a proper vent. Other issues:

- Oil vapors from mechanical pumps may need to be vented to prevent back-streaming into the region being evacuated.

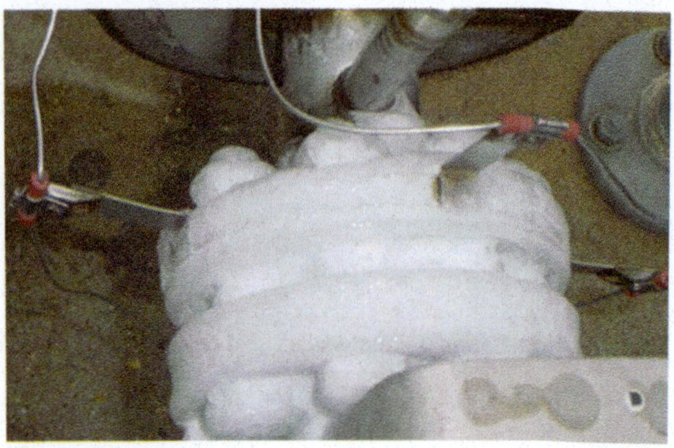

Fig. 6.18 Ice built-up on an uninsulated fitting [4]

- Leaks in an evacuating system can result in system being contaminated with air.
- Be aware that a vacuum pump with ballast valve can result in a combustible mixture within the pump or its exhaust. This can be prevented by replacing the air supplied to the ballast valve with nitrogen.

The insulating vacuum protecting cryogenic systems requires maintenance as well. A buildup in pressure within the insulating vacuum indicates a problem with pumping. The system may need to be purged to remove unwanted H_2 or other gases.

Maintenance. All materials and components should be subject to a comprehensive inspection and be quality-controlled. Maintenance program must be approved and sustained as needed. Inspection should occur at least annually. Maintained by qualified personnel according to approved procedures.

Inspection should be performed only if equipment is made safe for such maintenance:

- Lubrication
- Instrumentation calibration
- Cleaning and painting
- Operational verification of relief and check valves
- Replacement of filter elements
- Repair or replacement of
- Damaged or faulty components
- Components subject to wear (seals, seats, bearings)

6.7.2 Operations

Actions, techniques and considerations that have proved valuable in past hydrogen work are introduced by topic.

Policy. Operations include normal operating procedures, performing modifications, repairs or decommissioning of a system. Policies, safety analysis, and protocols directing work should be written. Prior to introducing hydrogen, or "wetting" a new system/facility all procedures should be reviewed, and participating staff trained and certified. Training should be regularly conducted to ensure continual safe use of H_2. Any system modification, if different from maintenance should be subject to review. Don't innovate without review. Emergency planning should extend to all contingencies in an operation.

Successful operations pursue a variety of considerations to minimize the severity and consequences of unplanned equipment failure or operator error. This includes using only the amount of hydrogen needed. Control of access to the work area to account for who is present in a potentially hazardous area and to keep nonessential personnel away from harm. Keep the work area ship shape, apply use good housekeeping practices to keep egress paths open. Be rigorous about use of personnel protection, even if working conditions are hot or unpleasant. All operations that would put personnel in intimate contact with hydrogen containing components or component at cryogenic temperatures should provide written instructions directing which PPE is applicable and when it is to be used. Ensure H_2 gas and H_2 fire detection systems are functioning properly and that alarms and warning devices are working.

Situational Awareness. An approach to promoting overall safety is for personnel to apply the control volume concept to "situational awareness" and be aware of controls that make their system safe. In addition, to carefully reviewing the status of system sensors before entering a hydrogen work area, one can apply a "Sherlock Holmes" mind set:

- Observation, do things sound as they should? If not, back away, until the situation is understood.
- Logic, check likely leak points, is the appearance ok. This might be observing frost buildup where it should not occur, or discoloration of metal suggesting heat, or visual distortion of light suggesting flames or plumes, etc.
- Evidence of insipient failure, unusual corrosion, or condensation, etc.
- Be aware of emergency planning, are egress routes clear?

Hazardous Operations. Certain operations are inherently more hazardous, for example; storage and transfers in which large amounts of hydrogen are off-loaded, or disposal of large quantities of hydrogen by flaring. These are activities that shouldn't be performed without an observer at a safe vantage point and preparations to immediately call for help and initiate rescue. The "buddy system" system is the term given for this strategy. The term does foster erroneous notions like coworkers sharing a task. Nor should personnel assigned observer status proceed to multitask, taking attention away from the hazardous work.

Dangerous Weather Conditions. It is sensible to monitor weather conditions for operations that are conducted out of doors. Heavy rain, excessive wind and electrical storm activity are criteria for cancelling operations. There are lightning detection systems that can notify when strikes are occurring 5, 10 or 15 miles away. Electrical storms can induce dangerous potentials in conductive infrastructure such as cross-country pipelines capable of electrocution, let alone supporting an electrical discharge in a flammable mixture. Work protocols should cover the steps needed for a rapid shutdown and time allowances included. The proper action is to cancel or discontinue operations at the approach of an electrical storms. Put the system in a safe state: Isolate, vent, and purge to remove H_2 or air.

Good Housekeeping Practices. Weeds or similar combustibles are not permitted within 25 ft of LH_2 equipment (29CFR 1910.103, Hydrogen). Access and evacuation routes are to be kept clear of equipment. In an indoor setting conductive and non-sparking floors are to be kept clean of dirt.

Reduction of Ignition Sources. Personnel must recognize unwelcome sources of ignition that might inadvertently show up in their work environment. Examples include:

- Control smoking, open flames, welding, use of mechanical tools, remove equipment that doesn't belong; generators, automobiles etc.

- Apply preventative action; install lightning protection, ensure continuity in bonding and grounding connections, use conductive machinery belts, and use explosion-proof or purged enclosures for electrical equipment that is likely to be exposed to flammable mixture (a code requirement).

But, no matter how vigilant the effort, always assume an ignition source is present!

Storage and Transfer Operations. These operations are inherently dangerous. At the start operators should notify surrounding workers to avoid the area, then establish area control. Storage and transfer operational areas should be kept clear of nonessential personnel. The buddy system should be in force. During transfers operators should be on heightened alert for leaks. When leaks occur operations should cease, the system made safe, repairs made before resuming the operation.

Vent Fires. Vent fires are a common occurrence and lightning a common cause. With an inert gas system, the procedure for extinguishing vent fire is to:

- Add inert gas flow, such as He,
- Stop H_2 flow,
- Continue inert gas flow until metal cools,
- Restart H_2 venting,
- Stop inert gas flow.

Emergency Procedures. Advanced planning is required to truly handle emergencies. Local emergency services should be informed of hazards and required mitigations. Inviting emergency service personnel to see facilities in advance is recommended. It is also advisable that they be informed in advance of hazardous operations. The operators of hydrogen systems should be kept aware of the onset of dangerous weather conditions and rapid decision making in place to order shutdown activities in a timely fashion. Primary aim of emergency procedures is to protect life and prevent injury. Responses to common situations are summarized:

- Leaks: The primary danger from a leak or spill is fire. Recall, hydrogen flames are difficult to detect and likely are practically invisible in daylight. Because of the low emissivity of the flames human senses cannot prevent inadvertent entry into flames.
  - Leak Procedures: Exclude people and vehicles from leak area. Isolate source, vent, purge, and repair. During this time avoid ignition sources.
  - Do not deliberately flare a leak. If it is dissipating without harm, let it do so.
- Fire Procedures: Let released hydrogen burn until supply can be cut off. To put it out before the release is stopped is to invite worse accumulation and reignition.
  - Use water spray or fog to protect system elements and stop fire from spreading.
  - Do not spray water on vent systems or relief valves. In cryogenic systems this can plug vent systems and set the stage for overpressures, rupture and fireballs.
  - In some instances, for example portable storage, it is desirable to remove a burning vessel from nearby vessels, but only if it can be done safely.

- Avoid Asphyxiation: Areas near spills may be oxygen deficient. While released hydrogen may rapidly rise the contents of damaged inert gas systems will not. Having an oxygen monitoring capability is important.
- Tank entry (H_2, N_2, He) requires applying confined space considerations, some of which are:
  - Having an entry plan, with emergency plans before any work is done,
  - Ensuring a fresh air supply,
  - Monitoring the atmosphere inside the tank, and
  - Following standard safety precautions.
- Medical response: Is to quickly and safely remove the injured from a danger zone. Rendering aid is best performed by trained medical personnel. Call for help.

6.7.3 Hazard Assessment

There are a variety uses for performing hazards assessments. See also Chap. 8. A list of hazard assessment uses might include:

- Design improvement,
- Safety evaluation and failure analysis,
- Formal assessments to communicate to management (obtain "buy-in" and oversight),
- Providing written documentation as a reference for operations,
- And, post mortem assessment, following a failure or accident.

In the development of a complex hydrogen system or a facility assessment, techniques can be applied to all stages of hydrogen work, for example; the initial concept, design reviews, operations, modifications to design or operation, and finally decommissioning.

Beyond identifying obvious concerns, these notes will hopefully provide some additional insight into how to look at hydrogen systems and approach hazard assessment.

First, consider that hydrogen releases differ from other fuels due to the extent of interaction with surroundings:

- A leak at a point can grow into a cloud affecting a large area with many potential combustion hazards,
- Large flammability range and low MIE promote interaction of released hydrogen with just about any conceivable ignition source,
- The kinetics of hydrogen combustion are significantly more rapid than those of most other combustible gases,
- Cryogenic issues are distinct from gaseous issues (releases can begin heavier-than-air and warm to become buoyant) but include all of the issues associated with gaseous hydrogen,

- Hydrogen combustion processes are often intertwined with the geometry of the physical system and surroundings, such that flame acceleration and development of dangerous overpressures can occur and occur rapidly.

For these reasons the complexity of phenomena can obscure identification of hazards, and it is recommended that hazard assessment should be done by personnel with hydrogen experience.

The primary accomplishments expected of safety assessment might be noted as:

- Systematically and objectively identify hazards
- Examine consequences
- Evaluate risk
- Identify mitigations
- Document and communicate
- Provide mechanism for control of hazards.

There are a variety of methodologies have been used for addressing hazards issues, each with features that may be more productive than others depending on the nature of the work (see Engineers workbook). Some examples include:

- Cause Consequence Analysis
- Energy Flow/Barrier Analysis
- Event Tree Analysis
- Fault Tree Analysis
- Risk Assessment Matrix Success Tree Analysis.

From the author's experience the complex interaction of hydrogen within a system and with the environment when released call for an approach that uses both *inductive* and *deductive* reasoning in understanding hydrogen hazards. With inductive thinking, reasoning proceeds from individual cases to a general conclusion and in the case of hazards induces the consequences of an event forwardly (bottom-up). A useful analysis of this kind is a failure modes and effects analysis (FMEA). When conducted thoroughly every system component is examined for the outcome that would occur from possible failures. For example, a valve might be evaluated for the consequences of failing open, failing closed, or leaking. The consequences for each state are evaluated to see if they directly result in hazard or induce other component failure that results in hazards. This approach addresses hazards that come from system malfunction but doesn't address more global hazards which might be deduced or determined be reasoning from the general to the specific. One can ask general questions like' how might a fire hazard arise, or what if the operator commits a certain error. These two approaches to searching for hazards must also account for design or operational actions that mitigate hazards. Ultimately, to be useful the combined analysis should show either how hazardous consequences are reduced, or the likelihood of a hazard occurring is reduced or both. The control volume concept allows an analysis to focus on region of a system and on the elements that provide the control. If an FMEA approach is applied to the elements of the control volume of interest a hybrid approach is created that uses both inductive and deductive components together. The effect of a given component failure can be assessed against all the controls on the volume to evaluate hazards (see HHAP). Simplifications in analysis can be realized by only looking at the control volumes affected by hydrogen and by applying the analysis to a group of similar components.

In aerospace systems hazard scenarios are typically not well characterized (system may be unique) with detailed information on leaks and the potential combustions associated with them, therefore hydrogen hazards assessment has many qualitative aspects. This is addressed with qualitative risk assessment matrices into which the probability of ignition and combustion outcome is evaluated by expert opinion and assessment is grouped into categories (not possible, remote, possible, probable, and highly probable. Consequences are considered in general areas such as personnel safety, program objectives, function capability in which the risk is qualitatively assessed; A negligible, B Marginal, C Critical, and D Catastrophic. The two different sorts of information are arrayed in a matrix form that shows a topography of increasing risk on one axis and increasing consequences on another axis as an aid to decision making. Ultimately health exposure issues, materials issues and environmental concerns must also be included for assessment.

Most institutions provide their own approaches to hazard assessment. The elements described here may be found in their protocols, and if not, they may be incorporated. The primary points are hydrogen hazards are identified by systematic evaluation of the elements that make up a system or facility. The control volume concept focusses analysis on volumes into which hydrogen is released and looks at the controls used to manage the safety.

6.8 Facilities

6.8.1 Safety of Facilities with Hydrogen Systems

A facility is a place, or piece of equipment provided for a particular purpose, and with our interests, either being modified to accommodate a hydrogen activity or being built expressly to support hydrogen work. Codes generally identify *hydrogen storage systems as hydrogen systems*. However, when pursuing an application, the user will likely view the storage as separate from the hardware at the point of use for the hydrogen. Code direction regarding the point of use system are limited to certain specific applications (examples include fuel cells, electrolysis, thermal spraying, use of special atmospheres, etc.), or will specify to follow the guidelines provided by the manufacturer of the point of use system. Applications not specifically identified by code, that is "custom" or developmental systems are expected to be treated as a hydrogen system (storage) on the basis of the quantities of hydrogen processed within the system. All of the issues and consideration previously reviewed now apply. In some senses a hydrogen facility is just an expanded hydrogen system, that just consider the system concept in Fig. 6.8, now housed in or associated with some structure and particular location. The goals of facility safety involve, most importantly, the protection of the public, workers, and the environment. Secondary concerns are the equipment of value, the importance of the mission or hydrogen work, and public perception.

Facility operations incorporate safety policy which is based upon careful review of hydrogen hazards, use of written protocols, hazards communications and training. There is a specific focus on hazards arising from unplanned release of hydrogen and combustion, the hazards of pressure systems, proper consideration of materials and temperature issues, and protection of employees. Goals are to have safety considered in design and construction, for the implementation to be as foolproof as possible, to incorporate safety and hazard analyses, input from designers, operators, and safety engineers. Facility operations will provide good maintenance and activities are conducted with safety committee oversight.

6.8.2 General Facility Guidelines

Guidelines. When chartered to contemplate a hydrogen facility the many of the topics of interest have already been introduced: Facility siting, buildings and special rooms (test chambers if you're running a laboratory), piping and storage, and disposal. They are now reconsidered in the context of a facility.

General. The layout of the facility should separate the point of use area from storage and transfer activity in a safe fashion. If the facility features walls, enclosed areas or rooms the structure must meet code specifications that include where the hydrogen system is located relative to the structure, the fire ratings for walls, other structural elements and exposures. The nature of the application and the process quantities involved will impact all electrical hardware, equipment, and installation around the immediate location of the hydrogen system. Safe disposal of the hydrogen can also affect the safety footprint of the facility. Other safety considerations include placarding, posting and labeling of areas, fluid lines and vessels. These common-sense actions must be done according to code requirements. Planning should include where hazardous operations will be conducted, what exclusion zones are needed to protect surroundings, how access control will be managed, and how emergency services will be coordinated.

Siting. When setting up a hydrogen facility the first quandary is usually where, and as in other endeavors it is *location*. The location decision is governed by the type of application and the quantity. Codes distinguish between gaseous hydrogen and the much denser liquid hydrogen. How systems using storage quantities less than 11 m^3 (400 scf) gaseous hydrogen or 150 L (39.7 gal.) of liquid hydrogen are managed is left to the discretion of the AHJ. To get a sense of how to look at applications, in terms of increasing hazard:

- Small quantities such as used in laboratory work, or any process that presents little possibility of mixing hydrogen with an oxidizer are consider minimal risk. The hazards are handling issues readily addressed with appropriate PPE.

- Storage applications or processes can involve large quantities of hydrogen. However, the hydrogen is isolated from any quantity of stored oxidizer leaving the only exposure to be in most situations the surrounding air. The vessel system is a code approved system, possessing redundant controls, and protected by a safe zone. The primary hazard is release of hydrogen into air and the potential for fire. The vessel design and the relief protection are such that external fire leading to vessel rupture and explosion is not considered a credible scenario.
- Activities in which hydrogen is used as a propellant, brought together with oxidizer for combustion, or processes that present a fair risk of release and mixing are considered the most risky and hazardous. The primary hazard is violent explosion and is addressed by large exclusion zones, barriers and remote operation.

The compounding issue is the quantity, and greater quantity translates into greater hazard. By illustration, launching a rocket fueled by a large amount of hydrogen and oxygen is an example of a very hazardous enterprise. Exclusion zones become very large, on the order of thousands of feet[1x] and involve issues of controlling air traffic. This thinking doesn't necessarily apply to hydrogen appliances whose designs minimize hazard for well-defined operates. Examples of appliances would be hydrogen automobiles, fork lifts, batteries, fuel cells or electrolysis systems.

The desirability of preferred location based upon convenience or function must always be subordinate to safety of the facility and surroundings. There are associated factors:

- Determining a safe distance from property boundaries, work areas with personnel, and exposures that are not easily mitigated (open grates to underground drainage lines, fuel depots, overhead powerlines etc.).
- Do the positions of existing facilities drive the choices? Do their activities pose a threat to the hydrogen work, or vice versa?
- How will hydrogen be transported to the point of use storage, by what public routes, and within the facility, by what private route? When supplier's vehicles approach, is there a safe holding area?
- From where will emergency services come if needed, and how long until they can respond?
- Do the quantities anticipated for use pose an environmental threat, or require disaster response communications with nearby communities?

Candidate locations are selected for comparison of attributes and exposures. Exclusion zones are dictated by code generally on the basis of quantity (quantity-distance), but also including specific exposure criteria. Barricades, dikes and impoundments are mitigations that can be used as protective features for equipment and facilities. The AHJ decides what regulations and codes apply and informs the review process that determines whether overall safety is adequately addressed.

Relevant Codes:

NFPA 2 Hydrogen Technologies Code

NFPA 55 Compressed Gases and Cryogenic Fluids Code

DOD 6055.9. Department of Defense Ammunition and Explosives Safety Manual (current revision).

Electrical. Component classifications must be evaluated for electrical hardware and components used in vicinity of gas connections. Several regions are specified for North American service:

- Electrical installations that either might be occasionally exposed to flammable mixtures or will as a matter of process be exposed to flammable mixtures (see 1, below)—within 0.9 m (3′) of a connection will require explosion proof rating, or an inert gas purged enclosure.
- Electrical hardware located within 0.9 (3′)–7.6 m (25)′ of a liquid hydrogen storage container.

Terminals points should not turn or loosen under use conditions and should be protected from foreign objects and contaminants to avoid shorting.

Other options for class I, Group B, Division 1 locations:

- Purged enclosures per NFPA 496
- Intrinsically safe
- Approved for Class I, Group C atmospheres.

Adequate bonding and grounding provided and verified. Careful attention is directed to reducing known ignition sources and limiting spark generation (consider static charge creation in flowing fluids and moving belts). Facilities must be adequately grounded and protected from lightning. Vent lines may benefit from a vent stack discharge rod. Facilities must have adequate ventilation and lighting.

Facility Rooms, Structures and locations. Test areas, control room, buildings and laboratory spaces are examined.

NFPA 70, National Electrical Code:

"Class I, Group B, Division 1" locations defined as within 3-ft from flammable mixture sources.

"Class I, Group B, Division 2" locations defined as within 25-ft of LH_2 storage, or potentially subject to flammable mixture exposure from accidental release of hydrogen.

Buildings. The structure and layout is set up to minimize personnel injury and facility damage in case of H_2 fire or explosion. One approach is to construct with lightweight, noncombustible materials according codes and regulations.[10] Other recommended aspects for a hydrogen facility are to avoid unventilated peaks in ceilings, use shatterproof glass or plastic in window frames, provide a 2 hr fire resistance rating for walls, floors, and ceilings, locate explosion venting in exterior walls or roof. Intrinsically safe heating can be accomplished with steam or hot water heating, or other indirect means.

Structures that contain H_2-wetted systems must be adequately ventilated. Whether by active or passive means, the ventilation rate should dilute the cumulative hydrogen

[10] 29CFR1910.103, Hydrogen.

leak or release to keep the concentration below 25% of LFL (1% by volume) or less. Operational requirements are to establish ventilation before bringing H_2 into the system and place safeguards to make sure ventilation is not lost during a power failure or emergency shutdown procedure. This may mean removing the hydrogen from the system. The ingestion of hydrogen into building air intakes must be prevented by following code direction on where hydrogen or vents may be located, or by active means, using sensors and controls to automatically close the air intake if H_2 is detected. Install H_2 sensors in building outlet vents if H_2 used inside. Suspended ceilings and inverted pockets are not desirable, but if they are unavoidable these air volumes require separate detection and ventilation. Detection and ventilation capabilities are of limited effectiveness on complex room geometries. Another option where a volume can be sealed to isolate it from hydrogen is to apply an internal positive air pressure. Hydrogen systems in enclosed spaces should be put in a safe mode when idle. This may be accomplished by purging hydrogen from the system or ensuring against air (or other oxidizer) entering the hydrogen system by maintaining a positive hydrogen pressure. This assumes the enclosed space is adequately ventilated. It is important that facility operations and maintenance prevent contamination of purge and vent systems.

Control Rooms. When visual observation of hydrogen activity is necessary at location that does not meet the isolation requirements of code, a bunker or other reinforced structure that can protect personnel from the most severe credible event is advisable. Closed circuit cameras, mirrors, and sliding steel covers can augment protection of windows. Control conduits should be sealed against hydrogen intrusion and ventilation ducts closed off. Positive air pressure is applied to keep hydrogen from entering through small openings.

Laboratories. Work areas in which small inventories of hydrogen are handled (as specified by the AHJ, or code requirements (>45 scfm). Specialized equipment such as vent hood or ventilated cabinets that use high ventilation rates (50–150 air changes/h), detection systems combine with active controls to shut off hydrogen and apply purges, and special procedures combined with PPE are used to control hazards. Primary concerns are limited flash fires, small explosions (think of loud "pops"), and broken glassware.

Test Stands and Chambers. Experimental inventories can be increased by conducting tests remotely either in a concrete walled room or outside at an isolated area or test stand. Outdoor facilities might use a canopy or shelter and walls, but caution should be observed in the arrangement due to hydrogen's propensity to support accelerated flames with confinement. To avoid overpressures the enclosure should not have more than two walls set at right angles and there should be a vent region between the walls and roof or canopy.

Storage. The use of hydrogen storage systems that support facility work are installed and located according to code requirements. The maximum contained volume and the type of storage are used to determine the minimum distance to specified various exposures as specified by code. Some of those issues are the amount of clear space, the distance to stored flammables, the distance to nearby walls, the distance to public thoroughfares, the presence of overhead power lines or other fluid lines, etc. In the case of liquid hydrogen storage, there is an issue if a spill can enter into a nearby drain or culvert. Storage systems are typically located adjacent to the point of use, but isolated by the minimum allowed space according to the most stringent QD code requirement between the two applications. If liquid storage supplies a gaseous hydrogen application hardware designed to vaporize the liquid hydrogen and manage it at a desired pressure may be co-located with the storage. Another common approach is for the vaporized hydrogen to be transferred to interim gas storage at the desired use pressure. The gas system would be located to code.

Related Codes:
NFPA 2
NFPA 55.

Transfer Piping. Design must provide for adequate mounting (support, guides and anchors) as well as flexibility (expansion joints, loops, offsets) needed in the application. Changes in the pressurization within the line, especially with cryogenic fluids require relief protection for the piping. The piping must be electrical bonded across all joints and adequately grounded. Labeling (contents, flow direction) is required by code.

Related Codes:

- Design, fabricate, and test to ASME B31.3 and CGA G-5.4
- NFPA 2
- NFPA 55.

Facility Venting. Roof vents located 16 ft above roof can be used to vent up to 0.5 lb/s of hydrogen. Vents must be located so that H_2 does not get into building air intakes. Larger quantities of hydrogen should be disposed of by flaring through a flare stack or burn pond. The quantity and size of combustible cloud should be evaluated for thermal radiation from flame and the flare stack suitably isolated by an exclusion zone away from personnel, buildings, other facilities and exposures. Vent design must accommodate all weather conditions, wind loadings, have adequate lightning protection and charge dissipation against the onset of storms. Vents should not be located under or near electrical power lines or other elevated exposure.

Area surfaces and Roadways. Uninsulated LH_2 piping and components capable of forming liquid air should not be positioned over asphalt surfaces or roadways because of the potential exposure to liquid air and the subsequent formation of explosive compounds. Concrete channels should be inserted in the roadway where an elevated liquid hydrogen line is located.

Description of an enclosed hydrogen work area. As an example, a successful facility, a fuel cell vehicle garage, is described. Hydrogen vehicle work areas have been set up where the vehicle is position optimally for fresh humidified air, ventilation combined with detection draws air from high spots over the vehicle, and a nitrogen purge is plumbed for immediate access to car systems. Ground points for the vehicle and personnel are conveniently located to allow mobility around the vehicle. The vehicle stations are isolated on the garage floor from each other and sources of ignition. Standard tools, electrical outlet, workbench and wall are separated by a distance and well-defined painted regions. Operations are well defined. Vehicles are kept out in the open until they can be worked on. The on-board hydrogen supply is immediately isolated and supply lines purged. Unwanted hydrogen is disposed through roof top vents. Egress from the facility is clear and emergency procedures are simple, exit the work bay area to a safe holding area and allow emergency responders to safe the facility.

6.8.3 Protection of Hydrogen Systems

Protection of H Systems and Surroundings. Running a hydrogen facility entails a number of concerns among which are providing adequate separation of work activities, proper control of vehicles, controlling handling and location of flammables, proper isolation of power infrastructure, and for outdoor operations being able to accommodate harsh weather and react to sudden changes in weather.

Exclusion zones are used to achieve basic safety in a passive fashion as long as operational controls are in effect. An elaboration on the control volume notion is shown below (Fig. 6.19). The various elements that might be part of the overall protection of a component, system, facility, or hydrogen activity are depicted. The vessel walls incorporate design and code requirements. Leak detection may involve direct action of detection sensors, or inferred detection such as pressure drop, or other monitor of internal system logistics. Ventilation may be passive as in an open outdoor environment, or active by fans or other forced air flow. Up to this point intervention can happen to prevent combustion. Fire detection alerts operators to an ignited leak, and a safety hazard, and hopefully before more extensive damage can occur.

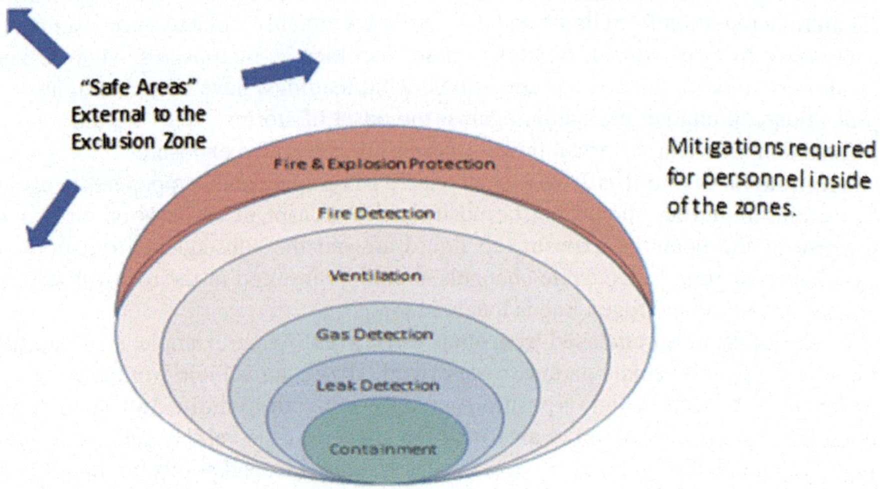

Fig. 6.19 Conceptual layers of mitigation [4]

The fire/explosion band represents an exclusion zone which is passive protection against the worst-case scenario; fire balls, explosion and shrapnel. Isolation can be modified, even reduced with the use of barriers. Walls, earthen berms, blast mats can be used successfully to reduce hazards [6]. Hazards that might come from other nearby facilities must also be included in an assessment of safety.

Against Unplanned Release. Site location preferences and necessary exclusion areas are based upon quantity-distance (QD) requirements. The minimum exclusion area is greater than or equal to the QD requirements and is driven by application and quantity:

Miscellaneous use (a one-time activity, or demonstration) is determined by site AHJ.

Laboratory scale operations (small quantities[11])

Non-propellant storage, or industry bulk storage

Propellant storage (on-board inventory, or hydrogen storage potentially subject to violence and direct mixing with large amounts of oxidizer

Must consider standard exposures [powerlines, drains, etc.]

- An exclusion area is set up with controls:
- Limit access to personnel with required training and proper protective equipment
- Ensure equipment is not an ignition source
- Operate according to approved procedures
- Personnel must use approved PPE
- Known hazards must be posted

Special Topic—Barricades, Dikes and Impoundments. Barricades can protect personnel and sensitive equipment from shrapnel and fragments. Protection of a liquid hydrogen supply by a blast mat is shown in Fig. 6.20. Earth mounds may also be used for protection, but their deployment must be based upon an understanding of shockwave behaviors [6]. Protective structures with confining walls must not provide confinement that can support transition to detonation.

Liquid hydrogen inadvertently released in amounts to create a spill may be contained with impoundments or dikes. Keeping the spill to a minimal footprint can limit the vaporization rate. Alternatively, crushed stone placed in the containment volume can add surface area for heat transfer to increase the liquid vaporization rate. Forcing the liquid into a smaller cross-sectional area pool can create a smaller combustion cloud but will take longer time to vaporize. The containment surfaces, or walls of dikes and impoundments should be kept as low as possible to ensure they do not provide confinement capable of supporting detonation in the flammable mixture that forms above the spill (Fig. 6.21).

[11] U. S. Occupational Safety and Health Administration regulation: $GH_2 < 11.3\ m^3$ (400 ft3), $LH_2 < 150$ L (39.6 gal).

Fig. 6.20 Blast mat used to protect a LH_2 trailer from shock and shrapnel [4]

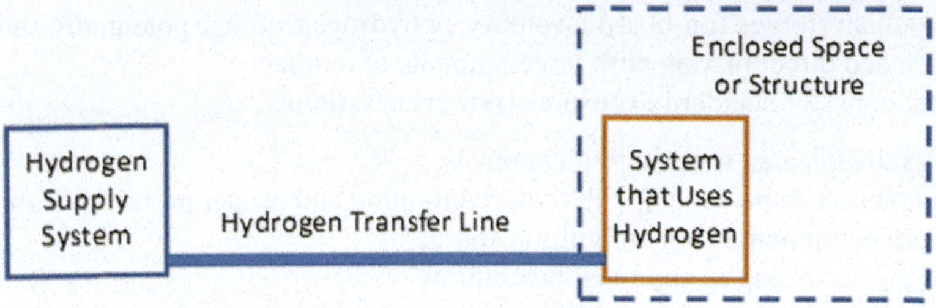

Fig. 6.21 Generic hydrogen facility/system [4]

6.9 Safety Checklist

A checklist,[12] contrived around a generic application, is noted to help identify considerations necessary to ensure a safe installation. The checklist is not intended to replace or provide guidance on compliance. Rather, it presents a concise table of critical safety measures compiled by some of the hydrogen industry's foremost safety experts.

It is a common practice for hydrogen applications to locate hydrogen supply systems outdoors for an indoor use as shown illustrated in the following diagram. The general principles in the checklist apply to all types and sizes of hydrogen systems.

Hydrogen safety, much like all flammable gas safety relies on five key considerations:

1. Recognize hazards and define mitigation measures (plan).
2. Ensure system integrity (keep the hydrogen in the system).
3. Provide proper ventilation to prevent accumulation (manage discharges).
4. Ensure that leaks are detected and isolated (detect and mitigate).

[12]Developed by the U.S. Department of Energy Hydrogen Safety Panel (For more information, visit: https://h2tools.org/).

5. Train personnel and ensure that hazards and mitigations are understood and that established work instructions are followed (manage operations).

The checklist is organized using these key considerations and is intended to assist with design and risk assessment. Examples are included to help users identify specific prevention techniques. While these considerations are fairly comprehensive, it is not possible to include all possible variables that need to be considered. The hazard analysis process should therefore include personnel who are familiar with applicable codes and standards in addition to team members that have expertise in the various technical aspects of the specific project.

Consideration	Approach	Examples of actions
Plan	Recognize hazards and define mitigation measures	• Identify risks such as flammability, toxicity, asphyxiates, reactive materials, etc. • Identify potential hazards from adjacent facilities and nearby activities • Address common failures of components such as fitting leaks, valve failure positions (open, closed, or last), valves leakage (through seat or external), instrumentation drifts or failures, control hardware and software failures, and power outages • Consider uncommon failures such as a check valve that does not check, relief valve stuck open, block valve stuck open or closed, and piping or equipment rupture • Consider excess flow valves/chokes to size of hydrogen leaks • Define countermeasures to protect people and property Follow applicable codes and standards
	Isolate hazards	• Store hydrogen outdoors as the preferred approach; store only small quantities indoors in well ventilated areas • Provide horizontal separation to prevent spreading hazards to/from other systems (especially safety systems that may be disabled), structures, and combustible materials • Avoid hazards caused be overhead trees, piping, power and control wiring, etc.
	Provide adequate access and lighting	Provide adequate access for activities including: • Operation, including deliveries • Maintenance • Emergency exit and response

(continued)

(continued)

Consideration	Approach	Examples of actions
Keep the hydrogen in the system	Design systems to withstand worst-case conditions	• Determine maximum credible pressure considering abnormal operation, mistakes made by operators, etc. Then design the system to *Contain* or *Relieve* the pressure • *Contain*: Design or select equipment, piping and instrumentation that are capable of maximum credible pressure using materials compatible with hydrogen service • *Relieve*: Provide relief devices that safely vent the hydrogen to prevent damaging overpressure conditions • Perform system pressure tests to verify integrity after initial construction, after maintenance, after bottle replacements, and before deliveries through transfer connections
	Protect systems	• Design systems to safely contain maximum expected pressure or provide pressure relief devices to protect against burst • Mount vessels and bottled gas cylinders securely • Consider that systems must operate and be maintained in severe weather and may experience earthquakes and flood water exposures • De-mobilize vehicles and carts before delivery transfers or operation • Protect against vehicle or accidental impact and vandalism • Post warning signs
	Size the storage appropriately for the service	• Avoid excess number of deliveries/change-outs if too small • Avoid unnecessary risk of a large release from an oversized system
	Provide hydrogen shutoff(s) for isolation	• Locate automatic fail-closed shutoff valves at critical points in the system such as storage exit, entry to buildings, inlets to test cells, etc.) to put the system in a safe state when a failure occurs • Consider redundant or backup controls • Install manual valves for maintenance and emergencies
	Prevent cross-contamination	• Prevent back-flow to other gas systems with check valves, pressure differential, etc.
Manage discharges	Safely discharge all process exhausts, relief valves, purges, and vents	• Discharge hydrogen outdoors or into a laboratory ventilation system that assures proper dilution • Direct discharges away from personnel and other hazards • Secure/restrain discharge piping

(continued)

(continued)

Consideration	Approach	Examples of actions
	Prevent buildup of combustible mixtures in enclosed spaces	• Do not locate equipment or piping joints/fittings in poorly ventilated rooms or enclosed spaces. Use only solid or welded tubing or piping in such areas • Provide sufficient ventilation and/or space for dilution • Avoid buildup of hydrogen under ceilings/roofs and other partly enclosed spaces
	Remove potential ignition sources from flammable spaces/zones	• Proper bonding and grounding of equipment • No open flames • No arcing/sparking devices, e.g. properly classified electrical equipment
Detect and mitigate	Leak detection and mitigation	• Provide detection and automatic shutdown/isolation if flammable mixtures present, particularly in enclosed spaces • Consider methods for manual or automatic in-process leak detection such as ability for isolated systems to hold pressure • Periodically check for leaks in the operating system
	Loss of forced ventilation indoors	• Automatically shut off supply of hydrogen when ventilation is not working
	Monitor the process and protect against faults	• Provide alarms for actions required by people, e.g., evacuation • Provide capability to automatically detect and mitigate safety-critical situations • Consider redundancy to detect and mitigate sensor or process control faults • Provide ability for the system to advance to a "safe state" if power failures or controller faults are experienced
	Fire detection and mitigation	• Appropriate fire protection (extinguishers, sprinklers, etc.) • Automatic shutdown and isolation if fire detected
Manage operations	Establish and document procedures	• Responsibilities for each of the parties involved • Operating procedures • Emergency procedures • Preventive maintenance schedules for equipment service, sensor calibrations, leak checks, etc. • Safe work practices such as lock-out/tag-out, hot work permits, and hydrogen line purging • Review and approval of design and procedural changes
Train personnel	• Material Safety Data Sheet (MSDS) awareness for hydrogen and other hazardous materials • Applicable procedures and work instructions for bottle change-out, deliveries, operation, maintenance, emergencies, and safe work practices	
Monitor	• Track incidents and near-misses and establish corrective actions • Monitor compliance to all procedures and work instructions	

6.10 Summary and Best Practices

Hydrogen is not inherently dangerous. It is how we use it where the balance between useful commodity and hazard lies. From this realization we proceed into how hydrogen work is conducted in the 21st century. The short list of best practices is:

- By planning and design that meticulously addresses possible faults,
- Careful review, hazard assessment, preparation of written protocols for operations are all prepared prior to actual use of hydrogen,
- Isolate hydrogen from oxidizers until the point of use,
- We isolate personnel and hardware we don't want to risk, unless we can adequately mitigate consequences,
- Instrumentation is used to monitor for hydrogen and controls applied to make a system inherently safe,
- Through training make personnel aware of potential hazards. The best safety-feature is a knowledgeable operating staff. Nearly 60 years ago R. B. Scott[13] observed on safety, "…no installation using liquid hydrogen can be made "idiot-proof". There is no substitute for intelligence." This thought can be extended to the use of hydrogen under whatever circumstances.

Bibliography

Looking for specific physical data or practice information? Some useful sources are:

- The U.S. Department of Energy Hydrogen Practice website http://www.h2incidents.org. In addition to best practice information there is a hydrogen Incident Reporting and Lessons Learned Database.
- NFPA 2 Hydrogen Technologies Code: http://www.nfpa.org.
- Edeskuty, Frederick J. and Walter F. Stewart; Safety in the Handling of Cryogenic Fluids. Plenum Press, New York, 1996.
- Berman, Marshall; Hydrogen Behavior and Nuclear Reactor Safety. In: Recent Developments in Hydrogen Technology, Volume II; Williamson Jr., K. D. and Frederick J. Edeskuty; Eds. CRC Press, Inc. Boca Raton, FL, 1986.
- Goldberg, B., Everhart, K., et al. System Engineering "Toolbox" for Design-Oriented Engineers, NASA Reference Publication 1358, NASA, Marshall Space Flight Center, December 1994.
- AIAA. Guide to Safety of Hydrogen and Hydrogen Systems. ANSI/AIAA G-095-2004. American Institute of Aeronautics and Astronautics, 1801 Alexander Bell Drive, Reston, Virginia 20191, 2004.

[13] R. B. Scott was manager of the National Bureau of Standards, Boulder Laboratories in 1963 [1]

References

1. D.B. Chelton, Safety in the use of liquid hydrogen, in *Technology and Uses of Liquid Hydrogen*, ed. by R.B. Scott (The Macmillan Company, New York, 1964)
2. R.H. McCarty, *Selected Properties of Hydrogen (Engineering Design Data)* (U. S. Department of Commerce, National Bureau of Standards, Washington, DC, 1981)
3. R.G. Zalosh, *Comparative Analysis of Hydrogen Fires and Explosion Incidents (C00-44442-2)* (Factory Mutual Research Corp, Norwood, MA, 1978)
4. NASA, *NASA Hydrogen Safety Course* (NASA Johnson Space Center, White Sands Test Facility, Las Cruces, New Mexico, USA, 2016)
5. W.U. Notardonato, *Ground Operations Demonstration Unit Form Liquid Hydrogen Initial Test Results* (NASA Kennedy Space Center, FL, 2015)
6. Department of the Army, *Structures to Resist the Effects of Accidental Explosions (Technical Manual TMS-1300)* (Department of the Army, Washington, DC, 1969)
7. R.D. Witcofski, *Liquid Hydrogen Spill Experiments: An Overview of Completed Simulated Storage Tank Rupture Experiments and Proposed Pipe Rupture Experiments* (Aeronautical Systems Division, NASA Langley Research Center, Norfolk, VA, 1981)
8. P.W. Hooker, *Experimental Releases of Liquid Hydrogen* (Health and Safety Laboratory, Buxton, Derbyshire, UK, 2011)

Chapter 7
Liquefied Natural Gas (LNG) Safety

Abstract The use of liquefied natural gas (LNG) to move natural gas efficiently from producer to consumer is a major industry. This chapter discusses the hazards associated with LNG. The properties of LNG are covered and unique hazards such as stratification and rollover are discussed. More common hazards such as flammability, pressure rise and oxygen deficiency are described. The phenomenon of rapid phase transition is also covered. Reference to existing codes and standards is emphasized. A list of best practices is included.

7.1 Example Accident

In 1971 at La Spezia, Italy a Liquefied Natural Gas (LNG) storage tank suddenly released much of its contents to atmosphere. Over the course of the next 3 h, a total of 300,000 m^3 of LNG vapor were vented. The peak boil-off rate during this event was as much as 250 times the normal value for that tank. Fortunately, while some of the vapor drifted onto public lands, no ignition occurred and no one was injured [1].

The cause of this violent, unexpected venting is a phenomenon known as rollover. Rollover, described in more detail below, occurs when density stratification occurs in a multi-component liquid mixture. Eventually, the densities of the two layers become equal and spontaneous mixing of the different temperature layers takes place resulting an a very rapid boil off of the liquid.

Rollover is a safety hazard present in LNG and not present in single component cryogenic liquid storage such as in liquid helium dewars.

7.2 Introduction

Liquefied Natural Gas (LNG) is widely used to transport natural gas from production fields to consumers that cannot be connected via gas pipelines. Instead, the natural gas is transported in liquid form by sea. Typical transport connections include: Qatar to Japan and Europe, Trinidad and Tobago to Europe, Australia to China and Japan.

T. J. Peterson and J. G. Weisend II, *Cryogenic Safety*, International Cryogenics Monograph Series, https://doi.org/10.1007/978-3-030-16508-6_7

Liquefying the gas results in a volume reduction of 600 allowing for efficient transport without the use of large pressure vessels. Once the LNG has reached the consumer nation, it is generally converted into room temperature gas at the receiving terminal and them distributed via gas pipeline. In some areas, such as China, LNG is also distributed as a liquid via road tankers or, as in the case of Norway, distributed as a liquid by smaller sea tankers along the coast. The transport of LNG is a significant industry. In 2016, a total of 346.6 billion cubic meters of LNG were shipped worldwide [2].

LNG is also used in peak shaving plants. In these facilities, natural gas is liquefied and stored until energy needs rise. Then the LNG is turned into a gas and sent into a gas pipeline system for distribution. The East Ohio Gas Plant described in Chap. 1 was a peak shaving plant.

Recently, natural gas is being used as a marine fuel due to its much cleaner burning properties. In this application, the fuel is stored on the ship as LNG. A description of an LNG fueled ferry is given in reference [3].

Given its scale, the LNG industry has a strong safety record. The last accident with multiple fatalities occurred in Algeria in 2004 and at the time that was the most serious accident in 24 years, A more recent accident in 2014 in Washington State injured a number of plant employees but did not cause injury or damage outside the plant boundary. The safe use of LNG has resulted from the development of strict procedures and design codes governing the production and distribution of LNG. Research into topics relevant to LNG safety such as accident modeling [4], spills [5] and rollover [6] continues. The safe use of LNG as a marine fuel is a topic of a recent paper [7]. Continued awareness of LNG hazards and their mitigation is vital to safe operation. This chapter describes the properties of LNG and its associated hazards. Mitigations for these hazards and pointers to additional information are provided. Examples of LNG safety research are also discussed.

7.3 Properties of LNG

LNG is simply the liquid phase of natural gas. Natural gas is not a pure fluid but rather is a mixture of a number of compounds and elements. This complexity results in some of the more unusual hazards of LNG, such as rollover. The composition of natural gas varies somewhat from gas field to gas field but natural gas is generally more than 95% methane with smaller amounts of ethane, propane, nitrogen and carbon dioxide. The boiling point of LNG is 112.2 K. The density of LNG varies with composition and will change as a function of time as those components with higher boiling points boil off preferentially to others. This effect know as "weathering" can create hazards and must be taken into account during LNG system design and operations.

LNG is distinct from Liquefied Petroleum Gas or LPG that consists mainly of propane and butane. LPG has its own set of hazards and will not be addressed in this chapter.

Natural gas and LNG are odorless, colorless and nontoxic. An important point should be made about odor. The distinct smell that we associate with natural gas comes from the addition of the chemical mercaptan. However, mercaptan is only added as the gas enters the distribution pipeline. Thus, in an LNG plant or during LNG shipment gas leaks cannot be detected by smell.

LNG at 1 bar is less dense than water and thus any spills during sea transport will float on the water's surface. Natural gas at 1 bar and 300 K is also lighter than air and will disperse upwards away from a leak or spill. However, keep in mind that natural gas colder than room temperature can be heavier than air and may enter basements or sewers. This effect is part of the reason that East Ohio Gas Plant accident described in Chap. 1 was so destructive.

7.4 LNG Hazards and Mitigations

7.4.1 Hazards Common with Other Cryogenic Fluids

While it is easy to fixate of the flammability of LNG along with other, more exotic hazards such as rollover, it should be kept in mind that LNG is a cryogenic fluid. Thus, LNG also has hazards in common with nonflammable single component fluids such as liquid helium. Proper materials (Chap. 2) must be chosen for use in LNG designs. Interestingly, Invar is used in some LNG tank designs. The extreme cold of LNG can cause damage to eyes and tissue and the precautions described in Chap. 3 also apply to LNG.

Venting LNG can displace air and result in an Oxygen Deficient Hazard (ODH). The large volumes of LNG seen during transport and storage increase this risk. Thus, the analysis and mitigations described in Chap. 4 are relevant here as well. Keep in mind that the vapor from LNG is odorless so ODH monitors will likely be required. However, in the case of LNG, these monitors must be designed to work in a flammable atmosphere.

The large volume ratio between room temperature vapor and LNG at 1 bar means that, like other cryogenic fluids, pressure relief systems (Chap. 3) will be an important part of LNG system design. However, the flammability of the gas will dictate where it can be vented.

7.4.2 Flammability

The flammability range of LNG is 5–15%. This range is actually smaller than that seen with hydrogen (Chap. 6). Detailed design codes, both regulatory and industry standards, exist (see below) that describe the safe mitigation of the flammability hazard of LNG. The appropriate codes should be strictly applied to reduce this risk.

While it's best to refer to the codes and standards directly rather than to attempt a synopsis here, a few general statements can be made.

- Avoid any scenario that mixes LNG or natural gas with air, oxygen or any other oxidants. Many of the recorded accidents with LNG have resulted from accidentally creating a flammable mixture.
- Pay particular attention to properly purging production, transfer or storage equipment with an inert gas prior to introducing LNG. The 2014 Washington State accident was a direct result of improperly purged equipment.
- Install sufficient detection equipment so that gas leaks may be found quickly.
- Ensure that staff are well trained in procedures and safety requirements including knowing how to properly respond to an accident.

7.4.3 Rollover

Rollover is a phenomenon that can occur in multicomponent liquids such as LNG stored at 1 bar. Rollover is the result of the liquid stratifying into two different density layers. This situation is illustrated in Fig. 7.1. In the figure, there are two fluid layers, with the higher density layer on the bottom. Heat entering the storage tank sets up convective cells as shown. In the upper layer, the heat is convected to the free surface of the fluid where it causes evaporation. Heat entering the lower layer however, is blocked from the free surface of the fluid by the density difference and instead is deposited into the lower later warming the fluid in this layer.

The density differences required for this situation to occur are relatively low, 0.5–2%.

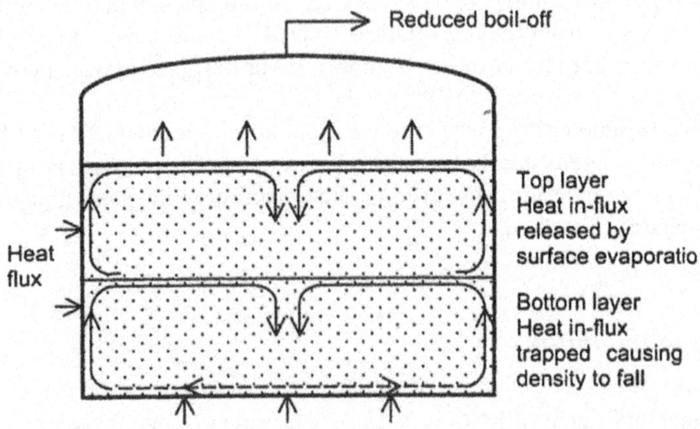

Fig. 7.1 Stratification of two different density LNG mixtures and associated heat flows. Without interventions, such a situation will result in rollover. From: *Stratification, Rollover and Handling of LNG, LPG and Other Cryogenic Liquid Mixtures*, R. Scurlock, Springer (2016) [1]

In the scenario shown in Fig. 7.1, as time goes on, the increasing temperature of the lower layer results in its density decreasing. At the same time, the evaporation of the lighter, lower boiling point, methane in the upper layer results in a larger fraction of ethane in the upper layer and thus density of the upper layer increases. Eventually, the densities of the two layers become very close and a sudden spontaneous mixing of the layers occurs. This mixing causes a rapid evaporation of the LNG due to the heat of mixing and the higher temperature of the lower layer. The result is a very fast increase in the boil off rate (BOR) and subsequent rise in pressure in the tank. In the worst case, this pressure rise would cause the tank to burst but even when sufficient pressure reliefs are installed in the system, a large amount of flammable natural gas will be released into the atmosphere.[1] Figure 7.2 shows the change in boil off rate seen during the La Spezia accident described at the start of the chapter.

Experimental studies have identified 2 different modes of boil off rate rises. "Mode 1 is a slower rise in BOR up to about 5–10 times that of the normal boil off rate while mode 2 is a sudden fast rise in BOR up to 100–200 times the normal rate" [1]. These studies were conducted with mixtures of liquid oxygen and liquid nitrogen and have indicated that any multicomponent stratified cryogenic liquid mixture can experience rollover. This, by the way, means that the storage of large amounts of liquid air as

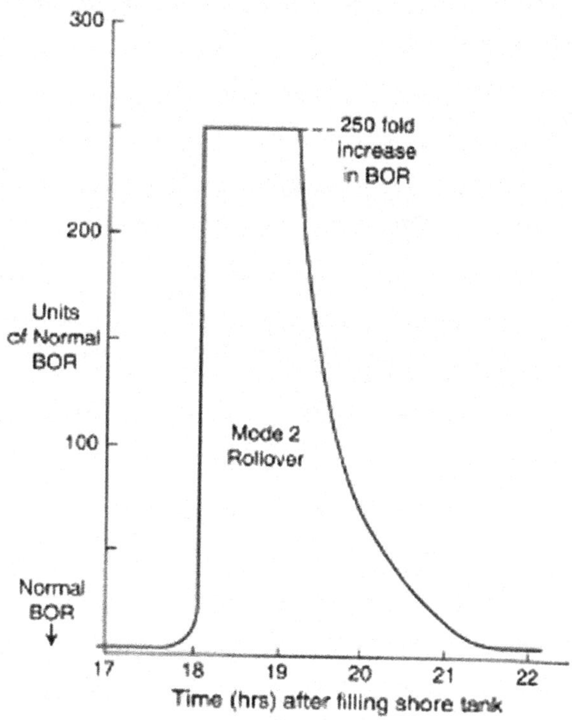

Fig. 7.2 Change in boil off rate (BOR) seen during the La Spezia accident. From: *Stratification, Rollover and Handling of LNG, LPG and Other Cryogenic Liquid Mixtures*, R. Scurlock, Springer (2016) [1]

[1]Even if the released natural gas does not catch fire, methane is a very significant greenhouse gas and its release into the atmosphere should be avoided.

part of proposed cryogenic energy storage systems [8] is at real risk of rollover and mitigations similar as those described below will have to be implemented.

A very detailed description of stratification, rollover and associated mitigations is given in reference [1].

7.4.4 Stratification

The initial necessary condition for rollover to take place is for the fluid mixture to stratify. A principal mechanism for this stratification to occur is to combine two slightly different density LNG mixtures together. Such combinations may occur when transferring LNG between ships and on shore storage tanks. Weathering refers to the phenomena of older LNG naturally become slightly higher in density due to the preferential boil off of the lighter, lower boiling point methane fraction. In the La Spezia accident, a "newer" less dense LNG mixture was loaded on top of the heavier more weathered mixture without any mixing this set up the stratified situation which resulted in the rollover.

Even if the LNG in a tank isn't stratified to start, there are mechanisms that may result in auto stratification. As described in reference [1], the presence of nitrogen, carbon dioxide or water in the LNG may through different mechanisms result in auto stratification and lead to roll over.

7.4.5 Mitigations

The most important mitigation to avoid rollover is to avoid stratification. It is particularly important to mix the LNG together (for example by as always filling the tanks from the bottom) when filling tanks. Other mitigations include suitable tank instrumentation to detect stratification and the capability to conduct mixing quickly if stratification is detected. Following the recommendations in reference [1] and in the applicable codes and standards can prevent stratification and rollover from becoming a problem.

7.4.6 Rapid Phase Transition

Since the majority of LNG transport is done by sea, spills of LNG either at sea or during loading or unloading activities will likely bring LNG into contact with sea water. LNG is less dense than water and any spilled LNG will float on the water's surface. Of course the temperature of the water is far above the boiling point of LNG and any spilled LNG will rapidly boil releasing natural gas. The evolved vapor is a combustion hazard, but there is an additional hazard related to the boiling process itself.

Under certain conditions the boiling LNG can undergo Rapid Phase Transition (RPT). The phenomenon can be thought of as explosive boiling. The explosion is physical rather than chemical i.e. there is no combustion but the shock waves generated from RPTs can be quite strong and lead to equipment damage or personal injury.

The video shown here (https://youtu.be/h-EY82cVKuA) illustrates the occurrence of RPTs in LNG. This footage is from some safety studies on LNG spills done by Gaz de France in the 1980s. Note that once the spill starts there is rapid continuous boiling punctuated by periodic larger explosions. These explosions are the rapid phase transitions. Note further that some of the RPTs are strong enough to shake the video camera located many meters away from the event.

What is the mechanism behind the rapid phase transition? The basic explanation is that when the LNG floats on the water surface, film boiling of the LNG occurs and a layer of gas vapor separates the LNG and water. Sometimes this layer will collapse putting the LNG and water into direct contact, which greatly increases the heat transfer between the water and LNG and thus the boil off rate resulting in a RPT. The problem is that explaining and predicting the occurrence of PRTs beyond this simple model is quite challenging.

Various studies [9, 10] have shown that the onset of RPT may be related to such factors as the percentage of methane in the LNG, the temperature of the water, the depth to which the LNG goes in the water and the spill rate. However, a consistent sold correlation between any of these factors and the start of a rapid phase transition has yet to be obtained. It appears to be true that small lab scale experiments do not reliably predict RPTs in full scale, real world situations.

In order to address this problem, a number of large-scale LNG spill experiments have been carried out over the years. Figure 7.3 shows the relationship between the spill rate and the energy released by an RPT of an experiment carried out at China Lake, CA in 1981. There appears to be a sudden increase in the energy release at the higher spill rates. Notice that the energy released from the RPTs can be more than the equivalent of 3 kg of TNT.

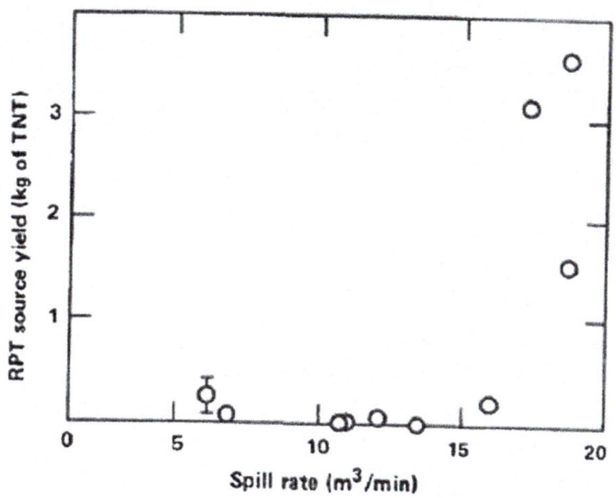

Fig. 7.3 Measured RPT energy release as a function of spill size [10]

Absent being able to predict conditions that will lead to RPTs, the best mitigation is to avoid spilling LNG into the water during transport and handling. The codes and standards discussed below address this issue for LNG.

Understanding Rapid Phase Transitions remains an important area of research in LNG safety [11].

7.5 Existing Codes, Standards and Publications

The LNG industry is heavily regulated. There are many codes, both required and advisory, which cover all aspects of LNG production, transport and storage. Strictly following the applicable codes and standards is one of the best ways to ensure safety in LNG systems.

The Society of International Gas Tanker and Terminal Operators (SIGTTO) exists to develop common standards and best practices for the safe transport and handling of LNG. Their website http://www.sigtto.org provides a wide range of useful information and their publications, particularly "Liquefied Gas Handling Principles on Ships and in Terminals, (LGHP4) 4th Edition" [12], are very useful.

Similarly, the Compressed Gas Association (https://www.cganet.com) has a useful publication on natural gas safety [13] that also includes some comments on LNG.

Examples of other applicable codes and publications [14–20] include:

- NFPA 59A "Production Storage and Handling of Liquefied Natural Gas"
- BCGA CP46 "The Storage of Cryogenic Flammable Fluids"
- The International Maritime Organization "International Code for the Construction and equipment of Ships Carrying Liquefied Gases in Bulk"
- U.S. CFR Title 49, Part 193: Liquefied Natural Gas Facilities: Federal Safety Standards (2006) [17]
- European Standard (EN) 1473: Installation and Equipment for Liquefied Natural Gas—Design of Onshore Installations (2016) [18]
- European Standard (EN) 13645: Installations and Equipment for Liquefied Natural Gas—Design Onshore Installations with A Storage Capacity Between 5 T and 200 T (2001)
- Environmental, Health, and Safety Guidelines: Liquefied Natural Gas Facilities, World Bank Group.

7.6 Best Practices

- Rigorously follow all applicable codes, standards and regulations regarding the production, transport and storage of LNG.
- Do not neglect typical cryogenic fluid hazards such as low temperatures, oxygen deficiency and overpressure when considering LNG safety.

- Be aware of unique hazards such as rollover associated with LNG being a multi-component fluid with components of different boiling points.
- Avoid stratification of stored LNG and have techniques and processes available, consistent with codes and standards, for mixing LNG if it starts to stratify.
- Take into account the phenomena of Rapid Phase Transitions in the case of LNG spills.
- Keep in mind that stratification and rollover may also be a hazard in other multi-component mixtures such as liquid air.

References

1. R. Scurlock, *Stratification, Rollover and Handling of LNG, LPG and Other Cryogenic Liquid Mixtures* (Springer, 2016)
2. BP Statistical Review of World Energy (2017)
3. M. Chorowski et al., LNG systems for natural gas propelled ships. Adv. Cryo. Engr. (IOP Conference Series: Materials Science and Engineering, vol. 101, 2015)
4. S. Rathnayaka, F. Khan, P. Amyotte, Accident modeling approach for safety assessment in an LNG processing facility. J. Loss Prev. Process. Ind. **25** (2012)
5. W.C. Ikealumba, H. Wu, Some recent advances in liquefied natural gas (LNG) production, spill, dispersion, and safety. Energy Fuels **28** (2014)
6. J.Q. Shi, C. Beduz, R.G. Scurlock, Numerical modeling and flow visualization of mixing of stratified layers and rollover in LNG. Cryogenics **33**(12) (1993)
7. L. Vandebroek, J. Berghmans, Safety aspects of the use of LNG for marine propulsion. Procedia Eng. **45** (2012)
8. R. Morgan, S. Nelmes, E. Gibson, G. Brett, Liquid air energy storage—analysis & first results from a pilot scale demonstration plant. Appl. Energy **137**, 845–853 (2015)
9. A. Luketa-Hanlin, A review of large-scale LNG spills: experiments and modeling. J. Hazard. Mater. **140** (2007)
10. R.P. Koopman, D.L. Ermak, Lessons learned from LNG safety research. J. Hazard. Mater. **A132** (2006)
11. E. Aursand, http://www.ntnutechzone.no/en/2017/02/lng-how-to-predict-an-explosive-boiling-accident/
12. *Liquefied Gas Handling Principles on Ships and in Terminals, (LGHP4),* 4th edn. (Witherby Seamanship, 2016)
13. *Guideline for Natural Gas Safe Handling and Operations* (CGA G-19, Compressed Gas Association, 2017)
14. NFPA 59A, *Production Storage and Handling of Liquefied Natural Gas* (National Fire Protection Association, 2016)
15. BCGA CP46, *The Storage of Cryogenic Flammable Fluids* (British Compressed Gas Association, 2016)
16. *International Code for the Construction and Equipment of Ships Carrying Liquefied Gases in Bulk* (The International Maritime Organization, 1993)
17. U.S. CFR Title 49, Part 193: Liquefied Natural Gas Facilities: Federal Safety Standards (2006)
18. European Standard (EN) 1473: Installation and Equipment for Liquefied Natural Gas—Design of Onshore Installations (2016)
19. European Standard (EN) 13645: Installations and Equipment for Liquefied Natural Gas—Design Onshore Installations with a Storage Capacity Between 5 T and 200 T (2001)
20. *Environmental, Health and Safety Guidelines: Liquefied Natural Gas Facilities* (World Bank Group, 2017)

Chapter 8
Approaches to Cryogenic Safety in Particle Accelerator Labs

Abstract A significant application of cryogenic technology is found in the cooling of superconducting magnets and superconducting radiofrequency cavities used in particle accelerators. This chapter describes the safety challenges found in such laboratories and the use of written policies, training, hazard analysis, reviews and lessons learned in creating a safe work environment in such facilities. A list of best practices is included.

8.1 Introduction and Challenges

Laboratories that employ particle accelerators for research are among the largest scientific facilities in the world and are frequently significant users of cryogenic technology. Examples of these laboratories include the European Center for Nuclear Research (CERN), The European Spallation Source (ESS) , The SLAC National Accelerator Lab, The Deutsches Elektronen-Synchrotron (DESY) and The Fermi National Accelerator Lab (Fermilab). A wide range of cryogenic applications is found in such facilities. These include: the use of superconducting magnets to bend and focus the particle beam and to provide background fields in particle detectors and sample environments; superconducting radiofrequency cavities that accelerate the charged particle beam, and vessels of cryogens such as liquid argon, liquid hydrogen and liquid xenon that act as detector mediums, moderators or scattering targets. Figures 8.1 and 8.2 show examples of large scale uses of cryogenics in particle accelerator labs.

Such laboratories present a number of challenges. Technical requirements based on scientific needs, such as thin windows on the low temperature vacuum system through which the particle beam passes or the need to use niobium as the material for superconducting radiofrequency cavities result in designs that can't be fully compliant to pressure vessel codes. The presence of ionizing radiation restricts the materials available for use and may limit the venting of gases to the environment as part of a pressure relief system. Since these facilities are generally very visible and publicly funded, there is typically a higher degree of oversight and stronger safety standards applied to them. Lastly, as expensive research facilities, these labs face

T. J. Peterson and J. G. Weisend II, *Cryogenic Safety*, International Cryogenics Monograph Series, https://doi.org/10.1007/978-3-030-16508-6_8

Fig. 8.1 A cutaway view of the LHC superconducting magnets in the CERN tunnel (*Courtesy* CERN)

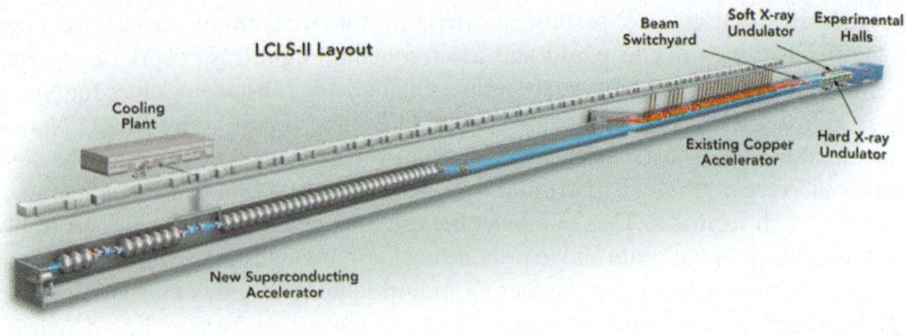

Fig. 8.2 View of the LCLS II project currently under construction at SLAC. Note the significant use of SRF cavities cooled to 2 K (*Courtesy* SLAC)

significant expectations from both users and funding agencies to be productive and reliable while maintaining safety.

Particle accelerator labs aren't the only ones that face these challenges. Facilities such as the ITER fusion project in France or the National High Magnetic Field Lab in the USA have similar issues as does NASA, ESA or other space programs.[1] How

[1] Due to weight restrictions, virtually no cryogenic container in a space system is built to the pressure vessel code.

these institutions meet these challenges in cryogenic safety is both interesting and potentially useful in other situations.

This chapter will describe various strategies that such laboratories use to meet these challenges while still maintaining safety in cryogenics and other systems.

8.2 Written Policies

One of the approaches that almost all these laboratories take is the creation of detailed, written safety rules and procedures. While these documents refer to industry standards and national regulations, they contain much more information. Creating these documents, which takes a significant amount of effort, has a number of advantages:

- The documents ensure that everybody working in the laboratory is using the same set of safety standards.
- This material provides an easily available safety reference not only for staff members but also for the large number of short term users that frequent these facilities.
- Rules and procedures that are unique to the laboratory in addition to national standards and codes can be described in these documents.
- The written policies generally provide a summary of the key points from applicable codes while referring to the codes for more detailed information. This increases the accessibility of the codes and standards to staff and users.
- The documents describe how systems that can't be designed to meet existing codes (such as the use of niobium at cryogenic temperatures in a pressure system—See Chap. 3) are reviewed and approved. The typical standard is that of "equivalent safety". If a component can't be designed to code then it must be shown through a combination of design, test and review that the component is as safe as it would be if it met the code.
- The documents can be reviewed by experts to ensure that they are correct and provide a safe environment. Once approved, the documents can be managed under a version control system so that only the most current version is available.

In the National Labs funded by the Department of Energy (DOE) in the USA, these policies are generally collected into an online Environmental Safety & Health manual. In other institutions such as CERN, the policies are written as distinct safety instructions for particular systems or technologies. Examples of the on-line ES&H manuals are given in references [1–3]. The CERN safety policy and rules, including instructions, are given reference [4] while reference [5] is a link to the safety policies at the US National High Magnetic field Laboratory.

An advantage of the on-line safety information produced by the DOE labs, CERN or the magnet lab is that these are frequently viewable by the general public. This information is carefully reviewed prior to publication and thus provides a valuable resource for various safety topics including cryogenic safety. Since these documents are frequently updated, one should take care that they are always referring to the most up to date version on line.

8.3 Hazard Analysis

Hazard analyses may follow various formats and procedures. Among these is the quantitative risk analysis for Oxygen Deficiency Hazards (Chap. 4) and the Job Hazard Analysis discussed in Chap. 1. We mention here two other methods that involve more qualitative assessments about hazards in a system. Failure Mode and Effects Analyses look at the impact of failures of individual valves and instruments. And a What-if Analysis considers the impacts of broader failures such as a power outage or cooling water system failure.

8.3.1 Failure Mode and Effects Analysis (FMEA)

Failure Mode and Effects Analysis (FMEA) provides a method for systematically stepping through the components of a system and assessing the severity of possible failures. Many forms of FMEA exist. For cryogenic systems, the Fermilab Environment, Safety, and Health Manual Chapter 5032, "Cryogenic System Review," [1] includes a suggested format for conducting an FMEA. This method consists of following a valve and instrument list or similar list of components and noting the following for each component:

- Failure or error
- Mode
- Hazard/Effect
- Hazard class
- Remarks/Recommendations

In considering the "failure or error" of each component, one asks the following questions about each type of failure each item in the valve and instrument list for a cryogenic system:

1. What happens if the item fails open or on when it should be closed or off?
2. What happens if the item fails closed or off when it should be open or on?
3. What happens if an instrument gives a false reading high?
4. What happens if an instrument gives a false reading low?

The answers to each of these questions result in a "hazard class'" which is a single-word conclusion: safe, marginal, or unsafe.

For example, if the failure of a valve to close results in inadvertent pressurization of a line, the result should be that the line is protected by a relief valve with adequate venting capacity, and the condition "safe". Table 8.1 illustrates the method for some elements of a superconducting RF system.

A conclusion of "unsafe" would imply the need for some kind of remediation to reduce the hazard. Thus, the FMEA is most effective as a design tool. One begins with a list of devices, such as a valve and instrument list and a process flow diagram,

Table 8.1 Excerpt from an FMEA. This selection shows considerations for "false readout" or "fails to operate" for some components of a superconducting RF system including RF cavities in helium vessels and a superconducting corrector magnet in the cryostat

Sensor description	Failure mode	Hazard/effect	Hazard class	Remarks and recommendations
Piezo actuators (fast tuners)	Fails to operate	Lose fast tuning	Safe	An operational problem for mitigation of microphonics
Stepper motors	Fails to operate	Lose slow tuning	Safe	An operational problem, cavity may not be tunable, so perhaps not operable
Helium vessel temperature sensors	False readout	False temperature information	Safe	Operational issue for cool-down or warm-up. Compare to adjacent helium vessels
Coupler tuner motors	Fails to operate	Unable to adjust tuning in test	Safe	Used in tests only, possible testing delay for repair
Superconducting magnet coil heater	Fails to operate	Slower magnet warmup	Safe	Cannot just locally warm magnet to kill persistent currents
Helium vessel heaters	Fails to operate	Inability to compensate lack of RF	Safe	Heaters on all 8 helium vessels provide redundancy
Magnetic sensors	False readout	Lack of local magnetic field info	Safe	Unable to compensate for imposed magnetic field, operational impact
Liquid level sensors	False readout	Loss of liquid level control	Safe	Redundant liquid level sensors provide backup

and works through the FMEA to determine if any additional engineering mitigations are required. The final FMEA then also serves as a quality assurance document for a system.

8.3.2 *What-If-Analysis*

The What-if analysis asks more general questions about events which affect the system more globally. For example, what if there is a power outage? What if a control system fails to communicate? In a manner similar to the FMEA, although the result will likely be an undesirable operational condition, the result should also be "safe" in that no personnel hazard results.

What-if differs from FMEA in considering events which may cause multiple failures. For example, while an FMEA asks the consequence of a valve failing in the closed position when requested to open, a What-if analysis may ask what happens if a power outage sends all the valves to their default positions (Table 8.2).

Similar to the FMEA, a conclusion of "unsafe" would imply the need for some kind of remediation to reduce the hazard. Thus, the What-if analysis should also be viewed as a design tool. The finished What-if analysis becomes a document which provides a record of some of the large-scale failures considered during the design process.

8.4 Reviews

Formal reviews of cryogenic equipment and systems to ensure that they are safe to operate are a common strategy used in large scientific facilities. This approach also works well in industry and smaller laboratories. The overarching goal of such reviews is to have an independent look at the system or component. The details of such reviews will vary by situation and institution but include everything from a peer review of an ODH analysis (Chap. 4) or pressure vessel design (Chap. 3) by a single engineer to a multiday safety review of an entire system by outside experts.

In order to be effective, safety reviews should have a well defined scope and charge. That is, what is the committee reviewing and what questions are the committee expected to answer? For example, the charge questions from the 2016 ESS Cryomodule Safety Review is given below:

1. "According to the hypothesis considered for the sizing of the pressure relief devices, is the current design protecting well the equipment (vacuum vessel, 2 K helium vessel, thermal shield circuit)?
2. Is the actual layout of the pressure relief devices venting in the tunnel acceptable from a safety point of view, especially regarding exposure to oxygen deficiency hazard and cold burns during maintenance phases?

Table 8.2 Excerpt from a what-if analysis for a superconducting RF system

What-if	Hazard/effect	Remarks and recommendations
Power outage-localized	Cryogenic distribution system shuts down with valves moving to their default positions. Most are spring-closed, air to open, so close. Cryogenic plant provides for its own protection independent of distribution system status. The two helium control valves on each cryomodule also move to failure mode position, which is "closed" for these valves. Liquid level decreases in cryomodule helium vessels	Safe but with program impact. Uninterruptable power supply (UPS) backup to control system for a short duration. Backup pneumatics to control valves for a short time via compressed gas storage but not necessary for safety
Power outage-site wide	Cryogenic distribution system and cryogenic plant shut down with valves moving to their default positions. Most are spring-closed, air to open, so close. The two helium control valves on each cryomodule also move to failure mode position, which is "closed" for these valves. Cryomodules slowly warm up	Safe but with program impact. Uninterruptable power supply (UPS) backup to control system for a short duration. Backup pneumatics to control valves by compressed gas storage. Helium inventory recovery with purifier compressor system using emergency generator
Loss of compressed air	All pneumatic control valves move to default position of actuators, in most cases closed positions	Safe but with program impact. Backup pneumatics to control valves for a while by compressed gas storage. System is protected by reliefs to accommodate dead-headed flow situations
Cryomodule loss of insulating vacuum	Expansion of helium in process circuits due to large heat load; process reliefs vent; helium inventory is lost	Safe condition: process piping reliefs are sized to accommodate loss of vacuum
Cavity beamline loss of vacuum	Expansion of helium in process circuits due to large heat load; process reliefs vent; helium inventory is lost	Safe condition: process piping reliefs are sized to accommodate loss of vacuum
Cryomodule internal circuit rupture	Loss of insulation vacuum to helium in cryomodule. Parallel plate relief on cryomodule vents helium to the tunnel	Safe condition. Relief system sized to protect helium vessel from pressure during internal circuit rupture. Potential ODH condition in Linac tunnel, personnel protected by ODH system
Cryoplant failure	Loss of cryogens to the Linac; Linac gradually warms up	Safe condition: cryomodule can safely warm up. Warm helium is safely returned to storage or vented from adequately sized relief system

3. Have all cryogenics and mechanical safety hazards been identified and its risks minimized, or otherwise mitigated?
4. Is the approach adopted against Oxygen Deficiency Hazard in the accelerator tunnel (engineering controls, organizational measures and personal protective equipment) acceptable from a safety point of view?" [6].

Inputs to such reviews may include: detailed technical descriptions of the equipment, hazards analysis (see above), results of quality and other verification tests, planned operating procedures and certificates of compliance to applicable codes.

Review committees should be made up of a mixture of experts in the technology being reviewed (e.g. cryogenics), safety experts, operations staff and people able to put the safety of the system being reviewed in context to the interfacing systems in the lab. Many institutions use a mixture of staff and external experts on such committees. Having external experts, particularly in the early stages of review and in large system reviews, brings in additional experience, may highlight poor assumptions and in general helps avoid "group think". Good judgment and independence are necessary qualifications for all review committee members.

In most institutions, committees do not approve a system as safe to operate but rather given an opinion to management on the safety of the system. Almost all reviews, result in recommendations for improvement or additional work. This, in fact, is one of the most valuable outcomes of a review. Below are some of the recommendations from the ESS Cryomodule Safety Review mentioned above.

1. "The sizing of the safety devices should be completed, extending the sizing rules to cover the vacuum vessel protection devices; the design of these devices should include the study of the escape path from the cold leak location, through the thermal shielding up to the position of the safety device on the vacuum vessel, taking into account the warm-up effect along the path to define the exhaust temperature of helium.
2. The ESS cryomodule team should elaborate a clear pressure test plan for the cryomodules, covering the complete lifecycle, from production of cavities, tests of the cryomodules, and testing of the cryomodules in the LINAC; this plan should identify the needs for pressure tests related to the pressure equipment category under PED; the pressure relief devices should be installed on the equipment after the pressure tests.
3. The hazard analysis should include a cryogenic operation scenario in which an erroneous opening of warm gas valves from the HP line feed the cryomodules with warm helium. This hazard should be analyzed, consequences assessed and mitigation measures found and whether it becomes a worst case.
4. Continue the study work on the collector as this is recognized to be a valuable solution to cope with the risk of ODH and cold burns to personnel from overpressure rupture of the burst disks; this study should be pursued with high priority as it may impact the design of the cryomodules and the integration of other equipment in the tunnel.
5. Access rules to the tunnel and in the vicinity of the cryomodules should be defined in relation to the various operating modes of the cryogenic system; access

during critical phases (for example cool-down and warm-up transients) should be forbidden. Adequate personal protective equipment against cryogenic hazards for personnel accessing the tunnel should be defined as a function of the operating modes or type of work" [7].

Having a system to formally track and close out recommendations is advised to gain the most benefit from safety reviews.

Examples given so far have been drawn from dedicated safety reviews. Such reviews are very important, but in a typical project many other reviews for example design reviews or project reviews are held. Since safety issues are best addressed early in the design process, questions regarding safety issues should be included in all stages of the design review, starting at the conceptual stage. Having a safety expert as a member of all design review committees is strongly recommended.

8.5 Lessons Learned and Near Misses

Preventing accidents by learning from previous accidents or near misses is a very effective way to improve safety. This approach is frequently referred to as using "lessons learned". Many safety rules and regulations are created in response to previous accidents and thus are driven by lessons learned. This book, with its example accidents and experience of its authors, is also strongly influenced by lessons learned.

Having a formal process to capture and distribute lessons learned can greatly add to safety. Many large scientific labs and industries have such a process.

Lessons learned can be drawn not only from accidents but also from "near misses", that is, events that didn't cause an accident but almost did. Near misses frequently illuminate a weakness in equipment or procedure that can cause a safety issue. Responding quickly to near misses can fix the underlying problem and prevent an accident.

The use of lessons learned requires that accidents and near misses be reported by workers. This can be hard to achieve. Any blaming of the messenger in the case of near misses should be avoided at all costs. Rather, a program that recognizes and rewards staff who point out safety issues or near misses is a much better approach.

Another problem that can occur with lessons learned is that the databases can be quite large and finding relevant examples may be difficult. Having appropriate search tools in a lessons learned database helps solve this problem.

References [8] and [9] are publically available lessons learned websites from the US Department of Energy and NASA. Reference [10] is a publically available lessons learned database for hydrogen systems. Figure 8.3 shows an example lesson learned in cryogenic safety from the NASA website. Reference [11] discusses some additional lessons learned from cryogenic accidents. John Jurns (the author of Chap. 5) writes a regular column on lessons learned in cryogenics for the publication *Cold Facts* [12].

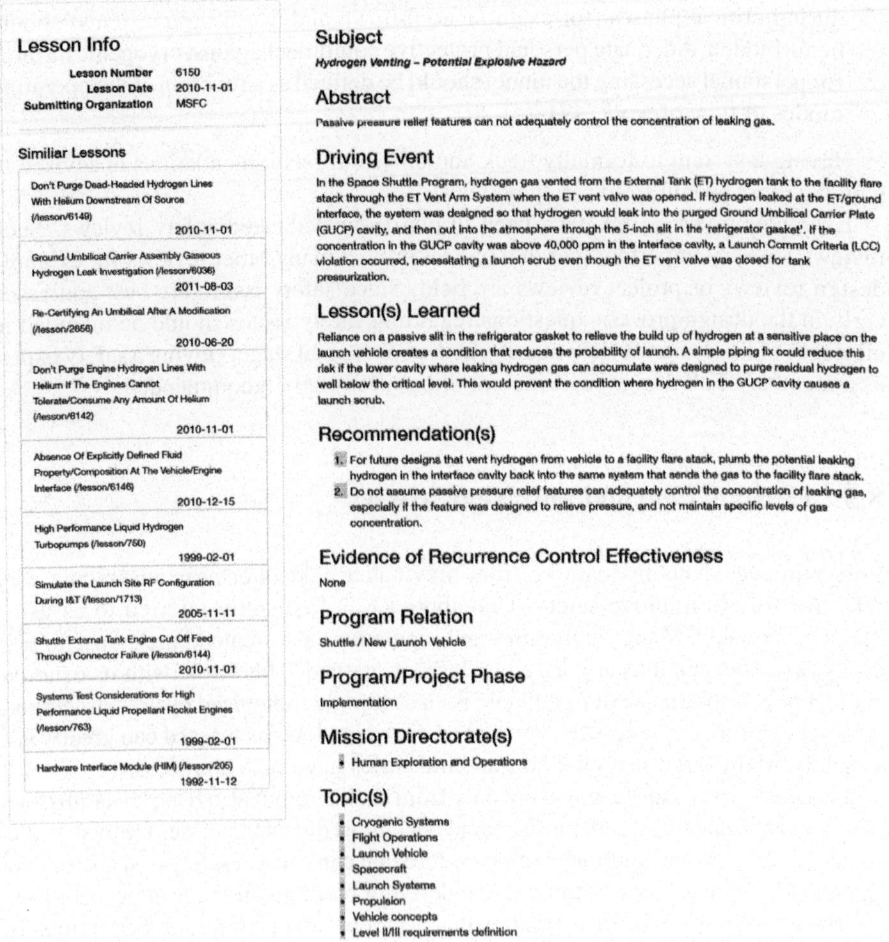

Lesson Info

Lesson Number	6150
Lesson Date	2010-11-01
Submitting Organization	MSFC

Similiar Lessons

Don't Purge Dead-Headed Hydrogen Lines With Helium Downstream Of Source (/lesson/6149) 2010-11-01

Ground Umbilical Carrier Assembly Gaseous Hydrogen Leak Investigation (/lesson/6036) 2011-08-03

Re-Certifying An Umbilical After A Modification (/lesson/2656) 2010-06-20

Don't Purge Engine Hydrogen Lines With Helium If The Engines Cannot Tolerate/Consume Any Amount Of Helium (/lesson/6142) 2010-11-01

Absence Of Explicitly Defined Fluid Property/Composition At The Vehicle/Engine Interface (/lesson/6146) 2010-12-15

High Performance Liquid Hydrogen Turbopumps (/lesson/750) 1999-02-01

Simulate the Launch Site RF Configuration During I&T (/lesson/1713) 2005-11-11

Shuttle External Tank Engine Cut Off Feed Through Connector Failure (/lesson/6144) 2010-11-01

Systems Test Considerations for High Performance Liquid Propellant Rocket Engines (/lesson/763) 1999-02-01

Hardware Interface Module (HIM) (/lesson/205) 1992-11-12

Subject

Hydrogen Venting – Potential Explosive Hazard

Abstract

Passive pressure relief features can not adequately control the concentration of leaking gas.

Driving Event

In the Space Shuttle Program, hydrogen gas vented from the External Tank (ET) hydrogen tank to the facility flare stack through the ET Vent Arm System when the ET vent valve was opened. If hydrogen leaked at the ET/ground interface, the system was designed so that hydrogen would leak into the purged Ground Umbilical Carrier Plate (GUCP) cavity, and then out into the atmosphere through the 5-inch slit in the 'refrigerator gasket'. If the concentration in the GUCP cavity was above 40,000 ppm in the interface cavity, a Launch Commit Criteria (LCC) violation occurred, necessitating a launch scrub even though the ET vent valve was closed for tank pressurization.

Lesson(s) Learned

Reliance on a passive slit in the refrigerator gasket to relieve the build up of hydrogen at a sensitive place on the launch vehicle creates a condition that reduces the probability of launch. A simple piping fix could reduce this risk if the lower cavity where leaking hydrogen gas can accumulate were designed to purge residual hydrogen to well below the critical level. This would prevent the condition where hydrogen in the GUCP cavity causes a launch scrub.

Recommendation(s)

1. For future designs that vent hydrogen from vehicle to a facility flare stack, plumb the potential leaking hydrogen in the interface cavity back into the same system that sends the gas to the facility flare stack.
2. Do not assume passive pressure relief features can adequately control the concentration of leaking gas, especially if the feature was designed to relieve pressure, and not maintain specific levels of gas concentration.

Evidence of Recurrence Control Effectiveness

None

Program Relation

Shuttle / New Launch Vehicle

Program/Project Phase

Implementation

Mission Directorate(s)

- Human Exploration and Operations

Topic(s)

- Cryogenic Systems
- Flight Operations
- Launch Vehicle
- Spacecraft
- Launch Systems
- Propulsion
- Vehicle concepts
- Level II/III requirements definition

Fig. 8.3 Example of a lesson learned in cryogenic safety from the public NASA lessons learned website

8.6 Best Practices

- Take advantage of the publicly available safety policies and rules developed by particle accelerator labs and other large scale scientific facilities.
- Conduct FMEA and "What If" hazard analyses.
- Carry out formal safety reviews of systems and include a safety component at all design review stages. Document these reviews including a list of recommendations for follow-up.
- Develop a system for capturing lessons learned from accidents and near misses.
- Take advantage of existing databases to learn from the broader cryogenic community.

References

1. *Fermilab Environment, Safety & Health*, Fermi National Accelerator Laboratory, http://eshq.fnal.gov/manuals/feshm/. Accessed 2018
2. *SLAC Environment, Safety & Health Manual*, SLAC National Accelerator Laboratory, http://www-group.slac.stanford.edu/esh/eshmanual/. Accessed 2018
3. T. Jefferson, *Environment, Safety & Health Manual*, National Accelerator Facility, https://www.jlab.org/ehs/ehsmanual/index.html. Accessed 2018
4. *CERN Safety Policy & Rules*, https://espace.cern.ch/Safety-Rules-Regulations/en/Pages/default.aspx. Accessed 2018
5. *Safety Training & Resources*, National High Magnetic Field Lab, https://nationalmaglab.org/user-resources/safety. Accessed 2018
6. Private Communication, Duy Phan, ESS, July 2016
7. V. Parma, J.G. Weisend II, A. Bernardes, B. Petersen, Y. Leclercq, D. Phan, P. Jacobsson, *Safety Review of the ESS Cryomodules—Review Committee Report*, ESS (2016)
8. *OPEXshare*, https://opexshare.doe.gov. Accessed 2018
9. *NASA Public Lessons Learned Database*, https://llis.nasa.gov. Accessed 2018
10. *H2 Tools Lessons Learned Database*, https://h2tools.org/lessons. Accessed 2019
11. F.J. Edeskuty, Accidents with cryogenic fluids and what we can learn from them. Adv. Cryo. Engr. **47**, AIP (2002)
12. *Cold Facts*, Cryogenic Society of America, https://www.cryogenicsociety.org/cold_facts/. Accessed 2018

Chapter 9
Summary and General Cryogenic Safety Guidelines

Abstract This chapter presents a brief summary of the text and lists a final set of general cryogenic safety guidelines.

9.1 Summary

The preceding chapters have shown that while there are many hazards associated with cryogenics, safe use of cryogenic systems is possible if generally well established safety mitigations and procedures are followed. The first step is to fully understand the hazards involved. Assumptions based on experience at room temperature frequently don't apply in the field of cryogenics. Ignoring this fact or neglecting to consider possible hazards is a root cause of many accidents. Poor planning, use of inappropriate materials and lack of proper communication are other frequent causes of accidents.

The scope and history of cryogenics have resulted in the development of many codes, standards and best practices that can lead to safety in cryogenics. These are described both in this text and in the references. Additional sources of information are given in the Appendix. Making use of these resources and of experts in cryogenics and cryogenic safety during the design, installation and operation of cryogenic systems is vital to safety.

The field of cryogenics is constantly evolving and with it cryogenic safety. New hazards may arise as new technologies develop and a better understanding of existing hazards and mitigations may occur. Thus, continuing education is important to stay current in cryogenic safety. Participating in the conferences and organizations listed in the Appendix is a very good way of maintaining an updated knowledge in cryogenic safety.

Throughout this work, we have used lists of best practices to summarize and highlight key points. As a final summary, General Guidelines for cryogenic safety include:

- Perform a formal hazard analysis on any cryogenic system or process prior to the start of work and design. Identify the hazards and how you will mitigate them. Ask "What If" questions. Keep in mind that equipment breaks, cryogenic fluids can turn rapidly into gas, valves leak or are improperly operated and vacuums can fail.

T. J. Peterson and J. G. Weisend II, *Cryogenic Safety*, International Cryogenics Monograph Series, https://doi.org/10.1007/978-3-030-16508-6_9

This analysis should be done regardless of the size or complexity of the cryogenic system.

- Design safety into equipment and processes from the beginning. Adding safety features at the end of the design process can be expensive, time consuming and may result in hazards being missed. Note that it is always better to remove a hazard via engineering design than to try to mitigate the hazard.
- Review, Review, Review. Everyone, even experts, miss things or make mistakes. Having the safety of your cryogenic system reviewed by other people, be they other members of your team, external consultants or formal review committees is key to improving the likelihood of a safe system.
- Even small changes to existing systems or equipment can result in a serious safety hazard. Have all changes in equipment or operations, no matter how trivial, reviewed by cryogenic and safety experts.
- Whenever working with cryogenic liquids or inert gases, no matter how small the amount, always consider the possibility that an Oxygen Deficiency Hazard (ODH) may exist. Either prove by analysis that such a hazard doesn't exist or apply appropriate design changes or mitigations to eliminate or reduce the hazard. ODH issues are particularly dangerous due to the large volume of gas evolved by even small amounts of cryogenic liquid and by the fact that in low enough oxygen concentrations, the first physiological symptom can be sudden unconsciousness, followed rapidly by coma and death.
- Take advantage of existing regulations and industry codes and standards to develop a safe cryogenic system.
- Never disable or remove safety devices, relief valves or operate systems outside their operating parameters.
- Only use materials at cryogenic temperatures that have been proven to work at those temperatures. Keep in mind during the hazard analysis that materials that nominally operate at room temperature (such as the outer walls of vacuum vessels) could reach cryogenic temperatures in certain failure modes.
- Ensure that everyone working with or around cryogenic systems, even casual or periodic users, have the appropriate level of training in cryogenic and ODH safety.
- Always use required Personal Protective Equipment and follow required operating procedures. Taking shortcuts frequently leads to accidents.
- Be aware that surfaces exposed to air and colder than 77 K will start to condense the air and that the condensate will be oxygen rich (due to the higher boiling point of oxygen). This can lead to an enhanced flammability hazard. Insulate surfaces to prevent air condensation and if unable to prevent it, manage carefully where the condensate collects and the materials with which it comes into contact. No ignition sources should be present.
- Working with hydrogen, oxygen and LNG presents additional hazards not present with inert cryogenic fluids and thus requires additional specialized knowledge. Make sure that such expertise is present in the designing, reviewing and operating of these systems.

- Multicomponent cryogenic fluids such as LNG and Liquid Air present additional hazards such a rollover. Ensure that these hazards are considered and mitigated during design and operation.
- Take advantage of existing information & expertise. There are significant resources available on cryogenic safety. These include books, pamphlets, research papers, codes & standards, websites and conferences. Cryogenic safety standards from other institutions such as Fermilab or CERN may also provide valuable information. The use of experts from other institutions as external reviewers or consultants may be beneficial.

Appendix: Additional Resources for Cryogenic Safety

In addition to the resources listed below, see the references at the ends of each chapter for further information specific to that chapter's topic.

Books & Monographs

1. F.J. Edeskuty, W.F. Stewart, *Safety in the Handling of Cryogenic Fluids* (Springer, 1996)
2. M.G., Zabetakis, *Safety with Cryogenic Fluids* (Plenum Press, 1967)
3. *Oxygen Deficient Atmospheres*, Compressed Gas Association Bulletin CGA SB-2, Compressed Gas Association (2014)
4. F.J. Edeskuty, M. Daugherty, Safety, in *The Handbook of Cryogenic Engineering*, ed. by J.G. Weisend II (Taylor & Francis, 1998)
5. T.M. Flynn, Safety with cryogenic systems, Chapter 10 in *Cryogenic Engineering* (Marcel Dekker, 1997)
6. *Cryogenic Safety Manual: A Guide to Good Practices*, British Cryogenics Council (2018)
7. F.J. Edeskuty, Accidents with cryogenic fluids and what we can learn from them. *Adv. Cryo. Engr.*, vol. 47, AIP (2002)
8. *Safe Handling of Cryogenic Liquids*, CGA P-12, Compressed Gas Association (2009)

T. J. Peterson and J. G. Weisend II, *Cryogenic Safety*, International Cryogenics Monograph Series, https://doi.org/10.1007/978-3-030-16508-6

Associations & Societies

1. Compressed Gas Association: A technical society covering all aspects of industrial gas production and use. They produce many useful safety guides including ones on oxygen use, design of pressure relief systems and Oxygen Deficiency Hazards, http://www.cganet.com
2. American Society of Mechanical Engineers: Producer of pressure vessel and other pressure safety codes and standards, https://www.asme.org
3. Cryogenic Societies: A number of technical societies exist covering aspects of cryogenics. All are dedicated to supporting the safe use of cryogenics and have a variety safety references and resources. Examples include:

   Cryogenic Society of America, http://www.cryogenicsociety.org
   British Cryogenics Council, http://bcryo.org.uk
   The Cryogenics Society of Europe, http://www.cryosoceurope.org
   The Cryogenics & Superconductivity of Japan, http://www.csj.or.jp/en/

Conferences

1. Cryogenic Operations Workshop—A biennial meeting (Even Years) concerning all aspects of cryogenic operations including safety. The link to the most recent meeting is given here: http://indico.ihep.ac.cn/event/7062/
2. International Technical Safety Forum—A periodic workshop covering safety (including cryogenic safety) at high energy physics and synchrotron light facilities. The link to the most recent meeting is given here: http://itsf2017.triumf.cab
3. Cryogenic Engineering Conference—A biennial (Odd Years) meeting concerning all aspects of cryogenics including safety. Papers are published in *Advances in Cryogenic Engineering*. The link to the most recent meeting is given here: http://www.cec-icmc.org
4. International Cryogenic Engineering Conference—A biennial (Even Years) meeting concerning all aspects of cryogenics including safety. The link to the most recent meeting is given here: http://www.icec27-icmc2018.org/

Laboratory Safety Chapters

A number of institutions have developed on-line safety manuals that describe their safety policies and requirements. Some of these are accessible to the public and may provide useful guidance. These policies and processes are subject to change so one should be careful to use the most recent online version.

1. Fermilab Environment, Safety & Health, Fermi National Accelerator Laboratory, http://eshq.fnal.gov/manuals/feshm/. Accessed 2018
2. SLAC Environment, Safety & Health Manual, SLAC National Accelerator Laboratory, http://www-group.slac.stanford.edu/esh/eshmanual/. Accessed 2018
3. Environment, Safety & Health Manual, Thomas Jefferson National Accelerator Facility, https://www.jlab.org/ehs/ehsmanual/index.html. Accessed 2018
4. "CERN Safety Policy & Rules", https://espace.cern.ch/Safety-Rules-Regulations/en/Pages/default.aspx. Accessed 2018
5. "Safety Training & Resources", National High Magnetic Field Lab, https://nationalmaglab.org/user-resources/safety. Accessed 2018

Lessons Learned Websites

These cover all aspects of safety not just cryogenic safety but contain search tools that allow relevant entries to be found.

1. "OPEXshare", https://opexshare.doe.gov. Accessed 2018
2. NASA Public Lessons Learned Database, https://llis.nasa.gov. Accessed 2018
3. H2Tools Lessons Learned Database (hydrogen safety), https://h2tools.org/lessons. Accessed 2019

Classes

1. "Cryogenic Engineering and Safety Course" by Cryoco LLC: http://www.cryocourses.com
2. "USPAS 2017 Cryogenics Course". On line course materials found at: http://uspas.fnal.gov/materials/17UCDavis/cryoeng.shtml
3. Cryogenic Safety Webinar, Cryogenic Society of America, https://www.cryogenicsociety.org/calendar/free_cryogenic_safety_webinars/

Index

T. J. Peterson and J. G. Weisend II, *Cryogenic Safety*, International Cryogenics Monograph Series, https://doi.org/10.1007/978-3-030-16508-6

The manufacturer's authorised representative in the EU is Springer Nature Customer Service Centre GmbH, Europaplatz 3, 69115 Heidelberg, Germany. If you have any concerns regarding our products, please contact ProductSafety@springernature.com

Printed and bound by CPI Group (UK) Ltd, Croydon, CR0 4YY
28/09/2026
02224077-0001